Bioepoxy/Clay Nanocomposites

Haipan Salam · Yu Dong

Bioepoxy/Clay Nanocomposites

Fabrication Optimisation, Properties and Modelling

Springer

Haipan Salam
Indonesia University of Education
Bandung, Indonesia

Yu Dong
School of Civil and Mechanical Engineering
Curtin University
Perth, WA, Australia

ISBN 978-981-16-7296-5 ISBN 978-981-16-7297-2 (eBook)
https://doi.org/10.1007/978-981-16-7297-2

This Springer imprint is published by the registered company Springer Nature Singapore Pte Ltd.
The registered company address is: 152 Beach Road, #21-01/04 Gateway East, Singapore 189721, Singapore

Preface

The latest development of epoxy nanocomposites has attracted industrial sectors to consider their product components in different fields such as automobile, aerospace, electronic devices and packaging materials. The enormous demand for these epoxy nanocomposite materials lies in the functionalisation of clay fillers within polymer matrices, which improves a wide range of properties of epoxy/clay nanocomposites. On the other hand, the environmental issue and the depletion of fossil fuel have obstructed further broad applications of epoxy/clay nanocomposites. As such, biodegradable polymers can be used as an alternative material to substitute for current conventional polymers extracted from fossil fuel. In particular, it has been shown that epoxidised soybean oil (ESO) has a great potential to be an alternative material to partially substitute for or even replace epoxy resins alone. In addition to its derivation from renewable sources, the functionalisation of ESO also increases biodegradability of bioepoxy polymer blends despite the reduction of mechanical and thermal properties. This book introduces an effective approach for manufacturing and characterising next-generation environmentally friendly bionanocomposites for the purpose of minimising the material use of conventional epoxy resins, one of least recyclable thermosetting materials in contemporary industries. The main aim is to substitute conventional epoxy with bio-based epoxy constituents in order to achieve bionanocomposite materials with good sustainability and eco-friendliness, as well as optimal mechanical properties in relation to the effects of bioepoxy content and clay type and content.

In this context, Chap. 1 reports recent progress in biopolymer/clay nanocomposites in view of material selection, fabrication, optimisation technique in material processing, as well as their assorted mechanical, thermal, biodegradable and barrier properties. Chapter 2 elaborates the material fabrication of both conventional epoxy/clay nanocomposites and bioepoxy/clay nanocomposites, and major characterisation techniques used for the holistic analysis of morphological structures, mechanical and thermal properties in both nanocomposite material systems.

Chapter 3 investigates the effect of material formulation and combined pre-mixing methods consisting of mechanical mixing, ultrasonication and centrifugation on mechanical and thermal properties of conventional epoxy/clay nanocomposites via in situ polymerisation. Both pristine and organomodified montmorillonite (MMT) clays with different organic modifiers, interlayer spacing values and clay contents were utilised to examine morphological structures, as well as mechanical and thermal properties of such nanocomposites. Chapter 4 reveals the impact of material formulation and processing parameters on mechanical properties of bioepoxy/clay nanocomposites via Taguchi Design of Experiments (DoEs) and Pareto analysis of variance (ANOVA) in order to identify the optimum combination of factors. Chapter 5 elaborates a clear relationship between optimum mechanical properties and morphological structures of bioepoxy/clay nanocomposites by means of Fourier Transform Infrared (FTIR) spectroscopy, X-ray diffraction (XRD) analysis, transmission electron microscopy (TEM) and scanning electron microscopy (SEM). Chapter 6 reports material properties of bionanocomposites such as tensile, impact and hardness properties, as well as thermal and biodegradable properties based on a final preferred combination of factors determined in Taguchi DoEs, which intends to become an essential guidance in manufacturing bioepoxy/clay nanocomposites. Chapter 7 successfully predicted mechanical properties of final optimal bioepoxy/clay nanocomposites by considering clay content, clay orientation and aspect ratio and ESO content. Voigt model and Reuss model, Hirsch model, Halpin-Tsai (H-T) model, modified H-T model, Hui-Shia model and laminate model were applied for the prediction of tensile modulus of bionanocomposites. On the other hand, the estimation of tensile strength of such bionanocomposites is based on Danusso-Tieghi (D-T) model, Nicolais-Nicodemo (N-N) model, Lu model and Turcsányi-Pukànszky-Tüdõs (T-P-T) model along with their comparisons with corresponding experimental data. Finally, Chap. 8 covers great potential applications of polymer nanocomposites as innovative advanced materials for the manufacture of automobile parts, active material packaging, tissue engineering and drug delivery.

The first author Haipan Salam kindly acknowledges Directorate General of Higher Education (DGHE), the Ministry of Education and Culture, Government of Republic Indonesia for awarding him a PhD scholarship to study at Curtin University, Perth, Australia. We are indebted to Mr. David Collier, Mr. Graeme Watson, and Mr. Andreas Viereckl at mechanical engineering laboratories, Ms. Elaine Miller at John de Laeter Centre and Mr. Ming Lim at material testing laboratory for their continued technical supports at Curtin University. Our sincere appreciation also goes to Associate Prof. Martin Saunders and Dr. Aaron Dodd at the Centre for Microscopy, Characterisation and Analysis (CMCA), The University of Western Australia for their kind assistance in XRD and TEM characterisation. A special mentioning is conveyed to Springer Nature book editorial team Loyola D'Silva, Saravanan Murugan and N. S. Pandian for their enormous efforts to warrant the successful publishing of this book.

We know that there is no perfection though we are striving for this in the book. Any valuable comments and critiques are welcomed by our leading researchers, scientists and engineers, particularly working in green composite materials and sustainable and recyclable composite materials. We anticipate that this book will become a useful reference material for better understanding of novel polymer bionanocomposite systems.

Perth, Australia

Haipan Salam
Yu Dong

Contents

Chapter 1
An Overview of Biopolymer/Clay Nanocomposites

1.1 Background

Conventional polymer nanocomposites use non-biodegradable materials that cannot be recycled and quite often become wasteful [1]. Moreover, the depletion of fossil fuel stock and plastic waste has become critical issues in recent years. Since the advent of twentieth century, the tremendous development of polymer materials at an industrial level has accelerated the increasing exploitation of raw materials extracted from fossil fuel or as by-products. Indeed, the total quantity of polymers produced from fossil fuel in the first decade of the twenty-first century was even greater than that combined in the past hundred years [2, 3]. Such a drastically increased production of plastic components inevitably raises environmental concerns to the public. Consequently, an urgent action need to be taken to utilise other alternative resources from natural polymers that can substitute for or even replace derived petrochemical polymers with natural renewable resources in possession of much better environmental sustainability and renewability, as well as cost-effectiveness.

Most recently, biopolymer materials are of a particular concern to scientists and engineers as a means for alternative sustainable and eco-friendly materials. Biopolymers can be classified into two types, namely bio-based polymers and biodegradable polymers [1, 4, 5]. Biodegradable polymers are described as polymer materials that can decline and degrade those properties when subjected to microorganisms, water and carbon dioxide [5]. Meanwhile, bio-based polymer can be readily biodegradable (e.g. plant oil based polymers, starch, polylactic acid (PLA)) or non-degradable. In terms of the fabrication of biopolymers, it can be categorised as renewable resource-based polymers, petroleum-based biodegradable polymers and biopolymers from mixed sources [4]. Among those, some research in bio-based polymers particularly focuses on bio-based thermosetting polymers such as epoxy, unsaturated polyester and phenol resins, as well as their composites. Researchers have had enormous interest in bio-based epoxies because of their similar triglyceride epoxy side groups, which are expected to be equivalent to those in conventional epoxy resins [6–9]. Petroleum-based epoxy resins with their excellent properties can be currently used in

H. Salam and Y. Dong, *Bioepoxy/Clay Nanocomposites*,
https://doi.org/10.1007/978-981-16-7297-2_1

high-performance applications such as aircraft components, filaments, pipes, tanks, pressure vessels and tools, which require the resin systems with improved mechanical properties and higher glass transition temperature (T_g) [10–13].

In addition, bioepoxy resins such as epoxidised plant oils are produced from derived plant oils. It can be manufactured through an epoxidation reaction from double bonds of plant oils with hydrogen peroxide in an acidic solution such as acetic acid or formic acid. Bioepoxy resins like epoxidised plant oils based on soybean and linseed are commercially available in large volumes with a reasonable price. Those epoxidised plant oil types are usually employed as lubricants and plasticisers for polyvinyl chloride (PVC) [6, 8, 14–19]. Among these, epoxidised soybean oil (ESO) has been widely investigated in the manufacture of polymer composites due to its abundance, inexpensiveness and high reactivity as co-monomers when combined with other monomer-based plant oils to form crosslinking networks [17, 20, 21].

Notwithstanding that ESO appears to be more attractive owing to its improvement in fracture toughness and functionality, high bioepoxy contents have been found to reduce tensile strength and T_g of bioepoxy blends due to the plasticisation effect, low reactivity and crosslink density when compared with conventional petroleum-based polymers such as bisphenol A epoxy resin [21, 22]. As such, in order to enhance the properties of bioepoxy blends, good compatibility and reactivity between resins and curing agents are essential to produce bioepoxy blends with a high crosslinking density.

Moreover, the addition of rigid nanofillers such as nanoclays, cellulose microfibres and basalt fibres can also improve mechanical and thermal properties of bioepoxy resins, as well as expand the applications of their composites [23–25]. As opposed to other fillers, nanoclay fillers have gained the great popularity due to their high surface areas within active platelet-like nanostructures. Toyota R&D laboratory researchers carried out the functionalisation of clay nanofillers in the early 1990s by using montmorillonite (MMT) clays as nanofillers to reinforce nylon-6. It was reported that the addition of no more than 5 wt% MMT clays yielded better mechanical and thermal properties than those of neat nylon-6 [26]. Recent studies have suggested that the addition of clay fillers can enhance mechanical and thermal properties of neat epoxy, as well as bio-based epoxy materials [22, 27–31]. Clay fillers have been widely reported by Alexandre and Dubois [10] to be attractive reinforcements because they can improve material performances at very low clay contents including the increases in strength and modulus, heat resistance and flame retardancy, as well as the decrease in gas permeability.

However, the properties of biopolymer/clay bionanocomposites are well known to be affected by the compatibility and reactivity of coupling agents including biopolymers, their catalysts and nanoclay fillers, as well as associated manufacturing methods and processing parameters [22, 28, 32–36]. The good compatibility between polymer matrices and their catalysts could improve the crosslinking density of resulting nanocomposites. Moreover, the compatibility between matrices and clay nanofillers can also influence the degree of clay wettability. The clay wettability between epoxy matrices is critical to achieve optimal mechanical properties of epoxy/clay nanocomposites. Obviously, the modification of clay interlayers could also enhance clay compatibility with better interfacial interactions between matrices and clay interlayers. Indeed, the good interfacial interaction between nanoclay fillers and polymer matrices leads to effective stress transfer and thermal shielding [11, 37–43].

Notwithstanding the compatibility of coupling agents, several processing parameters in material fabrication also affect clay wettability [44–47]. The used processing techniques and variables are crucial to nanoclay dispersion in polymer matrices. The homogenous clay distribution and favourable structure orientation of clay nanofillers within nanocomposites would lead to the significant improvements of their mechanical and thermal properties. Meanwhile, inhomogeneous nanoclay dispersion can be detrimental to mechanical properties of nanocomposites. Furthermore, the mechanical properties of nanocomposites are much more related to structural properties of matrices and fillers, which can be predicted by analytical modelling like using Halpin–Tsai model, Mori–Tanaka model and Tandon-Wang model or numerical simulations based on molecular dynamic (MD) simulation and finite element method (FEM) [48, 49]. For example, Brune and Bicerano [50] predicted the decreases of elastic modulus of nanocomposites with a sign of imperfect clay dispersion on polymer matrices by using Halpin–Tsai model. In addition, Miyagawa et al. [51] proposed combining Tandon-Weng and Halpin–Tsai equations with pseudo-inclusion model in predicting the Young's modulus of epoxy/clay nanocomposites with inhomogeneous clay dispersion.

On the other hand, among filler dispersion techniques, the combination of mechanical shear mixing and sonication is commonly employed in effectively dispersing clay fillers into polymer matrices [16, 22, 52–54]. In the mechanical shear mixing, clay aggregates can be broken up into fine clay particles by impellers and fluid shear forces. Atiemo-Obeng et al. [55] revealed that operation methods, states and properties of solid and liquid phases, unit operation, impeller geometry and mixing parameters are key factors in a solid–liquid mixing process.

The compatibility of petroleum-based epoxy, epoxidiser and curing agent for the preparation of bioepoxy/clay nanocomposites remains challenging. The major problem lies in the compatibility and optimum content of bio-based epoxy and curing agent required for material processing. Moreover, modified clay dispersion within the matrices and interfacial adhesion between clays and polymer matrices are worthwhile to be investigated in nanocomposite fabrication. A variety of processing parameters to manufacture bioepoxy/clay nanocomposites are highlighted in this study, which consist of shear mixing speed to achieve uniform clay dispersion within bioepoxy resins, the thermal-process effect, the duration, the pre-mixing speed of bioepoxy and clays, as well as the clay type.

1.2 Base Materials Used for Biopolymer/Clay Nanocomposites

1.2.1 Biopolymers

Biopolymers are developed as a solution to worldwide issues concerning the environment and the depletion of fossil fuels for polymeric materials owing to an increasing demand for environmentally sustainable biomaterials. The key factors to the production of biopolymers are (1) eco-friendly material resources with minimal wastes and reduction in CO_2 gas emission, (2) abundant resources and content maximisation of raw materials, (3) high cost-effectiveness on manufacturing processes with a wide range of applications and (4) light-weight materials with attractive material performances [1, 4, 5].

Currently, a wide variety of biopolymers have been produced commercially with different natural resources, ranging from soft rubber materials to hard biopolymers [56, 57], as listed in Table 1.1. Most of commercial biopolymers can be used to replace current petroleum-based polymers with the bioquantity up to 100% such as polyhydroxyalkanoates (PHA), PLA, cellulose, thermoplastics and derivatives starch, polyamide 11 (PA11), polyisoprene, polyester, polyethers and epoxy resins. As such, biopolymers can be classified into three major groups according to biopolymer types, namely biopolymers from 100% renewable sources, petroleum-based biodegradable polymers and biopolymers from mixed sources, as presented in Fig. 1.1 [4].

1.2.1.1 Renewable Polymers

Renewable biopolymers are extracted from natural resources. With some modifications, it can be used to replace or to substitute for current petroleum-based polymers. Plenty of biopolymers derived from natural resources have been produced commercially (Table 1.1). Among them, most biopolymers originate from plants and animal

Table 1.1 Commercial availability of biopolymeric production

Polymer	Biopolymer fraction	Biocontent (%)	Supplier
Polyhydroxyalkanoates (PHAs)	Whole polymer	100	Imperial Chemical Industries (ICI), Micromidas, Veolia, Denimer scientific
Polylactic acid (PLA)	L-Lactic acid	Up to 100	NatureWorks, Evonik, Corbion PURAC Biomaterials

(continued)

Table 1.1 (continued)

Polymer	Biopolymer fraction	Biocontent (%)	Supplier
Polytrimethylene terephthalate (PTT)	1,3-Propanediol	Up to 37	Dupont, BASF
Polyacrylic acid	3-Hydroxypropionic acid	10	Rossari Biotech
Polyethylene (PE)	Ethanol	10	Dow Chemical, Braskem
Polybutylene succinate (PBS)	Succinic acid	~54	Myriant, Revedia
Polyvinylchloride (PVC)	Ethanol	100	Solvay Indupa
Polyamide 11 (PA11)	Ricinoleic acid	100	Arkema
Cellulosics	Cellulose	Up to 100	Glycan Biotechnology
Thermoplastic starch (TPS) and starch derivatives	Starch	Up to 100	Novamont, Biograde Ltd. Biop, Biotec, Cardia Bioplastic, Cerestech, ENSO Plastic, Franplast, Grabio, Green dot, Kingfa
Polyisoprene	Isoprene	100	Ajonimoto Co. Inc., DuPont, Amyris, Glycos Biotechnology, Aemetis,
Aliphatic polyesters	1,3-Propanediol, succinic acid, fatty acids	Up to 100	NatureWorks, BioMatera, Perstop, Solvay, Novamont, Mitsui, Corbion and Dow
Polyethers	1,3-Propanediol, fatty acids	Up to 100	Dow, Arkema, DuPont, DuPont Tate & Lyle Bio Products
Polyurethane (PU)	Fatty acids	Up to 70	BASF, Bayer, Huntsman, DOW, and Yantai Wanhua
Epoxy resins	Triglycerides	Up to 100	Arkema, Polar Industries, Akcros Chemical, HOBUM oleochemicals, Hallstar, Dow, Entropyresin, Huntsman, Ecopoxy
Furan resins	Xylose	Up to 100	TransFuran
Polyamide 6/10	Ricinoleic acid	66	BASF, DuPont, EMS, PPA, Solvay, Evonik,
Polycaprolactone (PCL)	ε-caprolactone	Up to 100	Solvay, Union Carbide, Daicel

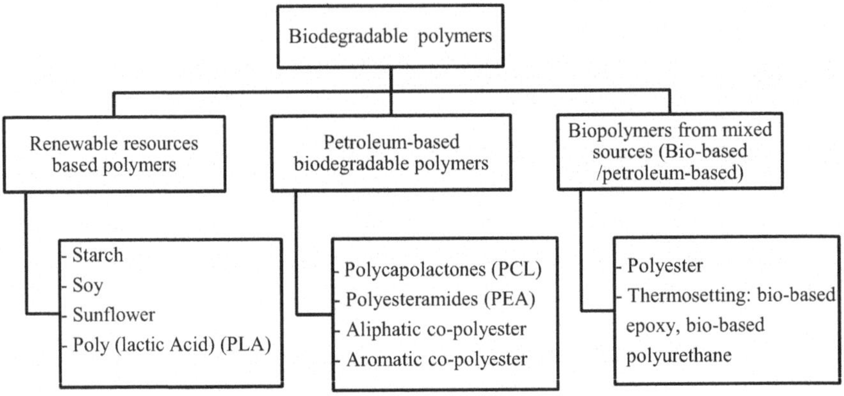

Fig. 1.1 Classification of biodegradable polymers [4]

gelatine (e.g. plant oils, thermoplastic starch and cellulose), as well as biopolymers with the conversion from bio-derived monomers in waste materials such as polylactides [4, 5].

Triglyceride based epoxy resins are modified and transformed from natural plant oils such as soybean oils, linseed oils, castrol oils, sunflower oils, etc. Those plant oils can be converted into epoxy monomers via the epoxidation reactions including Prilezhaev epoxidation, acidic ion exchange resin (AIER), chemo-enzymatic self-epoxidation, as well as metal-catalysed system reactions [58–60]. The epoxidation of plant oils is generally carried out by a conventional epoxidation process at the industrial level [59]. Among those plant oils, soybean and linseed oils are generally selected because of their high-unsaturated fatty acid contents to possibly produce epoxides with high functionality [14, 30, 59].

On the other hand, the compatibility between triglyceride-based epoxy and curing agents/initiators can affect mechanical and thermal properties of epoxy resins [59, 61]. Current progress in triglyceride-based epoxy cured by different initiators and curing agents are listed in Table 1.2. Park et al. [8] synthesised castrol oils and soybean oils via the homopolymerisation reaction with N-benzylpyrazinium hexaflouroantimonate (BPH) as a thermal latent catalyst. It has been found that T_g of epoxidised castrol oil (ECO)/BPH is 38 °C, which is higher than that of ESO/BPH at 24 °C.

Further, Park et al. [62] observed the effect of different cation latent catalysts such as N-benzylquinoxalinium hexafluoroantimonate (BQH) and BPH on mechanical and thermal properties of ECO. The results revealed that tensile strength and elongation at break for ECO/BQH were 47 and 13% higher, respectively as compared with those of ECO/BPH. This is attributed to different levels of crosslinking density. Meanwhile, T_g of ECO/BPH was found to be 8 °C higher when compared with that of ECO/BQH system. Additionally, Shibata [30] reported mechanical and thermal properties of ESO cured with different curing agents. The tensile strength and tensile modulus of cured ESO increased by 250 and 350%, respectively, with decreasing the ESO/tannic acid (TA) ratio (i.e. 1/0.8, 1/1, 1/1.2, 1/1.4). However, T_g of ESO cured by

Table 1.2 Triglyceride based epoxy resins cured by different initiators and curing agents

Epoxidised/plant oil	Catalyst			Curing agent	Curing process	Findings	References
	Promoter	Initiator/thermal cationic latent catalyst	Content (phr)				
Castrol oil	Amberlite IR-120	N-benyl-pyrazinium hexaflouroantimonate (BPH)	1	–	50–120 °C for 2 h	• Polymerisation initiated at > 60 °C • Tensile strength and elongation of ECO/BPH were 32 and 2% lower, respectively than those of BQH • T_g of ECO/BPH was 27% higher than that of ESO/BPH	[62]
ESO	Dimethyl benzyl amine	1,3-butanediol anhydrous	1.5	Maleic Anhydride	100 °C for 1.5 h; 120 °C for 2 h; 160 °C for 1 h	• Flexural modulus and shored hardness gained maximum values at 1:1 of epoxy: hardener ratio for about 432 MPa and 70 shore D • Maximum impact energy (24.1 J) was achieved with an epoxy/hardener ratio of 1:0.8	[63]

(continued)

Table 1.2 (continued)

Epoxidised/plant oil	Catalyst			Curing agent	Curing process	Findings	References
	Promoter	Initiator/thermal cationic latent catalyst	Content (phr)				
ESO	–	Pyridine	2%	Anhydride	-	• Curing agent content affected polymer properties • T_g increased with the increasing curing agent content that was influenced by increasing cross-link density • Storage modulus, T_g and cross-link density increased to maximum values at stoichiometric equivalent ratio • Izod impact and thermal stability showed constant values at curing agent contents of 0.7–1%	[15]

(continued)

Table 1.2 (continued)

Epoxidised/plant oil	Catalyst			Curing agent	Curing process	Findings	References
	Promoter	Initiator/thermal cationic latent catalyst	Content (phr)				
ESO	–	Tetraethylammonium bromide (TEAB)	0.3–0.8 phr	Methylhexahydrophthalic anhydride (MTHPA)	140 °C, with a variation of time for 1–3 h	• Crosslinking density and degree of conversion gradually increased with the increasing the catalyst content • T_g increased with increasing the catalyst content from 49.1 °C for 0.3 phr to 60.8 °C for 0.8 phr	[7]

(continued)

Table 1.2 (continued)

Epoxidised/plant oil	Catalyst			Curing agent	Curing process	Findings	References
	Promoter	Initiator/thermal cationic latent catalyst	Content (phr)				
Epoxidised cardanol	–	–	–	Isophorone diamine (IPDA) and diamine	120 °C for 5 h 170 °C for 30 min	• Crosslinking density of epoxidised cardanol cured with IPDA diamine was five times lower than that of diglycidyl ether of bisphenol A (DGEBA)/IPDA	[9]

(continued)

Table 1.2 (continued)

Epoxidised/plant oil	Catalyst			Curing agent	Curing process	Findings	References
	Promoter	Initiator/thermal cationic latent catalyst	Content (phr)				
ESO	2-ethyl-4-methylimidazole (EMI)			Maleopimaric acid (MPA)	–	• T_g increased up to 600% with increasing the curing agent ratio from 0.4 to 0.5 phr • Tensile strength, tensile modulus and elongation were improved by 250, 190 and 200% respectively, with 0.7 phr of MPA as compared to 0.5 phr of MPA	[64]
				Methylhexahydropthalic anhydride (MHHPA)	–	• Tensile strength and tensile modulus were improved by about 1550 and 3000% with increasing the curing agent content from 0.5 to 1 phr • The highest elongation at break was achieved with 0.7 phr of MHHPA	[64]

(continued)

Table 1.2 (continued)

Epoxidised/plant oil	Catalyst			Curing agent	Curing process	Findings	References
	Promoter	Initiator/thermal cationic latent catalyst	Content (phr)				
				Trimellitic anhydride (TMA)	–	• Tensile strength and tensile modulus were enhanced by 216 and 213%, respectively with increasing the curing agent content from 0.5 to 1 phr • Optimal elongation at break was achieved with additional 0.5 phr TMA • ESO with 0.7 phr TMA gained highest T_g as compared to other type anhydride-curing agent. T_g with 0.7 phr: TMA > MPA > MHHPA	[64]

1/1.4 TA ratio is 58 °C, which is 12 °C higher than that of ESO cured with a TA ratio of 1/0.8. In addition, the ESO cured with an anhydride curing agent like terpene-derived acid anhydride (TPAn) shows higher tensile properties and T_g when compared with ESO/TA cured at a similar ratio, which is 21.9 MPa, 745 MPa and 67.2 °C for tensile strength, tensile modulus and T_g accordingly. One of the examples for their commercial products includes ESO produced by Arkema (Vikoflex 7170), HallStar (Plasthall ESO and Paraplex G-60) and Akcros chemical (Lankroflex E2307) in addition to epoxidised linseed oils (ELOs). Commercially available epoxidised plant oils are commonly used as lubricants, poly (vinyl chloride) (PVC) additives, stabilisers, and bio-based renewable monomers to fabricate polyurethane rubbers [59].

On the other hand, starch-based polymers are also extracted from a natural resource (i.e. seeds, roots and tubes of different agro-resources such as corn, wheat, rice and potato) with good biodegradability [65]. The unique mechanical and thermal properties of starch have attracted much more attention in investigating material characteristics of starch. Liu et al. [66] reviewed the effect of thermal processing for thermoplastic starch on rheology, microstructures, and phase transformation in line with the processing conditions. It is noted that thermal processing can influence resulting microstructures and phase transition of starch-based polymers. Moreover, these microstructures and mechanical properties of starch materials, to a great extent, are affected by their manufacturing processing conditions and material formulations. A similar finding was also revealed by Babu et al. [5] with the effects of processing variables and material formulation on mechanical and thermal properties of starch edible films, and the effect of plasticisers and surfactants on their tensile properties [67, 68].

PLA is made from the bacterial fermentation of corn or sugar to form lactic acid. PLA, which is categorised as one of most popular biodegradable thermoplastics, can be used to replace conventional polymers such as polyethylene terephthalate (PET), polystyrene, and polycarbonate [69]. Physical, mechanical, and thermal properties of PLA are comparable to those of conventional thermoplastic polymers [70, 71]. Martin and Avérous [70] investigated the efficiency of plasticisers to PLA. It has been found that a variety of plasticisers such as glycerol, citric ester, polyethylene glycol (PEG), PEG monolaurate and oligomeric (lactic acid) can result in different properties of thermoplastic PLA. In particular, glycerol has been shown to be the least effective plasticiser among them. The addition of plasticisers to PLA can reduce T_g, as well as tensile modulus. Meanwhile, the crystallinity and elongation at break were detected to increase up to 29 and 200% at plasticiser contents of 10–20 wt%, respectively. PLA was synthesised for the first time in 1845 and started to be commercially available in the late 1990s, as pioneered by NatureWorks (previously known as Cargill-Dow LLC) [5, 60].

1.2.1.2 Petroleum-Based Biopolymers

Petroleum-based biopolymers like poly (ε-caprolactones) (PCLs), polyester amides (PEAs), aliphatic co-polyester and aromatic co-polyester are synthesised from monomers derived from fossil fuel resources and undergo the biodegradation at the end of use [4]. Biodegradable polyesters are classified as prevalent petroleum-based biopolymers with soft material behaviour even at ambient temperature [57]. PCL is semi-crystalline biodegradable polyester obtained by the ring-opening polymerisation of ε-caprolactones using a variety of anionic, cationic and co-ordination catalysts [72]. Besides, it is regarded as a hydrophobic polymeric material with low melting point and good solubility, which promotes its role as co-monomers to be combined with other monomer types via copolymerisation or crosslinking reaction [57, 72]. The other petroleum-based biopolymers are exemplified by PEA synthesised from diamines, diols, and diacids [73]. Nowadays, studies related to PEA have more focused on the biocompatibility of biopolymer blends [74], corresponding mechanical and thermal properties [75, 76], biodegradability [77] and their applications as biomaterials [73, 78].

On the other hand, aliphatic co-polyester and aromatic co-polyester are categorised as biodegradable polyesters. The commercial aromatic co-polyester (Biomax®, DuPont) is made from the PET polymerisation with the combination of aliphatic monomers like dimethyl glutarate and diethylene glycol. Meanwhile, aliphatic co-polyesters are obtained from the combination of diols (i.e. 1,2-ethanediol, 1,3-propanediol and 1,4-butanediol) and dicarboxylic acid. Currently, direct applications of biodegradable co-polyesters include commercial and consumer housewares, infant care, water bottles, small and major appliances and medical devices [56].

1.2.1.3 Biopolymers from Mixed Sources

The study of biopolymers from mixed sources has been the major subject in most research work. Biopolymer blends particularly aim to tailor biopolymer properties and to meet required and proposed applications. In general, biopolymer blends are deemed as the combined materials from renewable and biodegradable polymers, as well as petroleum-based polymers, which are either classified as biodegradable polymers or bio-based polymers. Moreover, to achieve final desired properties, biopolymers can be modified with the addition of co-monomers, plasticisers, fillers or reinforcements. Several parameters should be observed including compatibility [69] and interaction [79] between monomers and co-monomers. The compatibility of polymers can be tailored by adding compatibilisers, co-polymers or reactive diluent agents. Meanwhile, the interaction is also important in terms of molecular reaction and resulting structures to influence corresponding properties. Hence, the holistic investigation on the relationship between miscibility, structure, and properties is essential to be conducted to achieve final properties with specific applications [69].

The study of functionalisation of biodegradable polymers from mixed sources become popular from the point of view of their excellent biodegradability, reasonable cost, assorted material types, as well as social and environmental sustainabilities [2, 80]. Current developments of biopolymer blend are summarised in Table 1.3. It is mentioned that mechanical and thermal properties of biopolymer blends are

Table 1.3 Current progress in biodegradable polymer blends and associated objectives

Biopolymer blends	Outcomes	References
Starch/ polybutylene succinate (PBS) /Ionic liquid (IL) plasticiser	• Functionalised IL plasticiser improved the crystallinity of starch/PBS • Additional 0.3 wt% IL yielded the increases of 30 and 41% in tensile strength and elongation at break for starch/PBS, respectively, compared with starch/PBS/glycerol plasticiser	[81]
Gelatinised starch/PLA/glycerol	• Melting temperature T_m, T_g and crystallisation temperature T_c were reduced slightly with the addition glycerol • The tensile strength and elongation at break for gelatinised starch/PLA were improved by 17 and 45% as compared to those of pure starch/PLA/glycerol (10/90/3.85 wt%)	[82]
Thermoplastic starch/PCL	• Tensile modulus and tensile strength decreased for about 88 and 51% with increasing the glycerol content from 10 to 35 wt%	[83]
Starch/polyesteramide (PEA)	• Tensile modulus and tensile strength decreased for about 90 and 49%, respectively, with increasing the glycerol content (10–35 wt%)	[84]
PLA/polycarbonate (PC)	• Izod impact strength increased by 136% with increasing PBSL content up to 10 wt% • The optimum heat deflection temperature (HDT) was reached at 5 phr PBSL	[85]
PLA/ PBS	• PLA and PBS are immiscible in the molten state • Phase-separated structures were presented in PLA/PBS blends	[86]
PLA/polybutylene succinate (PBS) and PBS ionomer (PBSi)	• Storage modulus increased with increasing PBS and PBSi contents • PLA/PBSi blends were more compatible than PLA/PBS blends as predicted by the Flory–Huggins equation	[87]
PLA/polystyrene (PS)	• Incorporation of PLA in PS exhibited the destabilisation of PS • Partial miscibility of PS in PLA was detected by the existence of two different peak of T_g	[88]

(continued)

Table 1.3 (continued)

Biopolymer blends	Outcomes	References
PLA/PBS	• Increasing the PBS content enhanced the elongation at break for about 400% • Tensile strength and modulus were found to be 49 and 27% lower, respectively when compared with neat PLA	[89]
ESO/ diglycidyl ether of bisphenol F (DGEBF) and ELO/DGEBF	• The additional 30wt% ESO improved impact strength and fracture toughness of DGEBF resin by approximately 30 and 200%, respectively • The addition of ELO enhanced the storage modulus by 27% as compared to that of 30 wt% ESO	[90]
ELO/DGEBF	• Increasing ELO content reduced storage modulus, T_g and crosslinking density • The addition of 27 wt% ELO improved the impact strength by 350% as compared with neat DGEBF • Amine cured agent had lower crosslinking density as compared with anhydride • T_g of amine cured bioepoxy blends was lower than that of anhydride cured bioepoxy blends	[91]
ESO/DGEBA and ECO/DGEBA	• Multiple stages of curing reactions were confirmed for ESO/DGEBA/BPH resin blends • One stage of curing steps was found for ECO/DGEBA/BPH blends	[92]
ESO/DGEBA	• Joint strength increased with increasing the crosslink density • The addition of ESO could improve both impact and joint strengths	[6, 93]
ESO/DGEBA	• Polymerisation temperature of ESO/DGEBA/MTHPA increased with increasing the ESO content • T_g decreased with increasing the ESO content	[94]
ESO/DGEBA	• Tensile and flexural properties of ESO/DGEBA/ phthalic anhydride (PA) were reduced with increasing the amount of DGEBA resin • T_g increased with the addition of DGEBA resin	[15]

much worse than those of conventional polymers due to the plasticisation effect, low reactivity and lower crosslink density at high biodegradable contents [21, 22]. According to Khot et al. [14], increasing the renewable polymer content with respect to acrylated epoxidised soybean oil (AESO) in styrene biopolymers resulted in a decrease in both tensile strength and tensile modulus when compared to those of conventional styrene polymers.

In addition, Sahoo et al. [95] studied the effect of substitution of ESO into DGEBA resin on its mechanical and thermal properties, as well as morphological structures. Mechanical studies confirmed that tensile strength, tensile modulus and flexural modulus of bioepoxy blends decreased with the increasing ESO content. Whereas, tensile elongation and flexural strain were increased by about 26 and 56% with additional 30 wt% ESO. On the other hand, flexural strength and impact strength of ESO/DGEBA blends were enhanced by about 4 and 61% with the addition of 10 and 20 wt% ESO, respectively. Captured images by scanning electron microscopy (SEM) revealed an existing spherical domain of oil dispersed within DGEBA resin when 30 wt% ESO was used. Similarly, it was also found that the addition of biopolymers reduced tensile strength, flexural strength and flexural modulus [93] and hardness [96] of ESO/DGEBA blends, impact strength and T_g of ELO/DGEBA blends [91], T_g and storage modulus of ECO/DGEBA blends [8], as well as thermal properties of epoxidised vegetable oil (EVO)/DGEBA blends [92].

In order to improve the properties of bioepoxy blends, previous studies [13, 17, 97, 98] revealed that good compatibility and reactivity between resins and curing agents were essential in producing bioepoxy blends with high crosslinking densities. Shabeer et al. [21] reported an increase of 143% in the fracture toughness of 75 wt% biopolymers was influenced by the compatibility of DGEBA, ESO biopolymer blends and MHHPA as curing agents. Meanwhile, Espinoza-Perez et al. [61] found that the T_g of 30 wt% EVO was comparable to that of petroleum-based epoxy resin when curing agents such as MTHPA and MHHPA were employed. However, for those cured by amine curing agents (i.e. polyether diamine, cycloaliphatic polyamine and amine mixtures), its T_g was slightly lower than that of petroleum-based epoxy. Consequently, rigid fillers are functionalised to minimise the reduction in biopolymer properties.

1.2.2 Nanoclay Fillers

Many studies have attempted to alleviate poor mechanical and thermal properties with the incorporation of fillers. Several different types of fillers used in research studies include nanoclay fillers [16, 23, 31, 99], microfibrous celluloses [24, 100, 101] and basalt fibres [25]. Compared to other fillers, nanoclay fillers have gained the great popularity due to their attractive platelet-like nanostructures to be used for composite components with light weight structures. The unique structure and property of nanoclay fillers have resulted in the manufacture of many different polymer/nanoclay composites according to Alexandre and Dubois [10].

Nanoclay fillers are among the most popular fillers because clay minerals are obtained from naturally abundant resources that are readily used as potential fillers for composite materials. Clay minerals can also be developed as inorganic nanofillers for the synthesis of polymer nanocomposites owing to their economic value and environmental friendliness [23]. A small amount of nanoclay fillers can significantly improve the material performance of polymers as reinforcements and minimise the utilisation of polymers within composites. Besides, their layered aluminium and silicate structures facilitate the transformation of chemical properties and adjustment of compatibility with polymer matrices [102]. The inclusion of clay nanofillers enables not only to enhance mechanical properties of polymers [11, 103–113], but also to modify their physical properties by varying volume fraction, structure formation and particle size.

Nanoclay structures are associated with platelet-like nanoclays or layered silicates and tubular nanoclays in terms of filler shape. With the configuration of two tetrahedral sheets of silicate and one sheet layer of octahedral alumina, platelet-like nanoclays or phyllosilicates are formed, which include smectite, mica, vermiculite and chlorite. In particular, smectite clays are widely employed with further subcategories of MMT, saponite, hectorite, and vermiculite. The typical MMT clays are regarded as one of most effective nanofillers used in polymer/nanoclay composites due to their low material cost and easy intercalation and modification [102]. On the other hand, the fundamental structure of tubular nanoclays contains an aluminium hydroxide layer and a silicate hydroxide layer. They are also known as dioctahedral mineral with two different types of halloysite nanotubes (HNTs) and imogolite nanotubes (INTs). Notwithstanding their material role as clay minerals, these two types of tubular nanoclays resemble hollow tubular structures of carbon nanotubes (CNTs).

1.2.2.1 Montmorillonite (MMT) Nanoclays

MMT nanoclays are also known as 2:1 phyllosilicates and possess a layer of octahedral alumina and two linked tetrahedral silicate layers, illustrated in Fig. 1.2. A hexagonal symmetry is formed from one silicate atom and four joint oxygen atoms on the outer tetrahedral layer. Meanwhile, each aluminium atom is bonded with six oxygen atoms in the shape of an octahedron (i.e. a polyhedron with eight faces) on the octahedral layer with each phyllosilicate layer being linked by Van der Waals interactions through the interlayer galleries. The aluminium sheet can be easily replaced with a cation. Based on such a cation exchange method, Al^{3+} may be substituted for Mg^{2+} or Fe^{2+} to produce a negative charge on the layer that can be balanced by alkali cations such as Na^{+}, Rb^{+}, Cs^{+} or Li^{+}. A single phyllosilicate layer has a typical thickness about 1 nm while the thickness of its particle is in range from 30 nm to a few microns.

The commonly used MMT, saponite, hectorite and vermiculite are listed with their respective chemical structures in Table 1.4, which are based on their cation exchange capacity (CEC) to express the degree of cation substitution or substitution

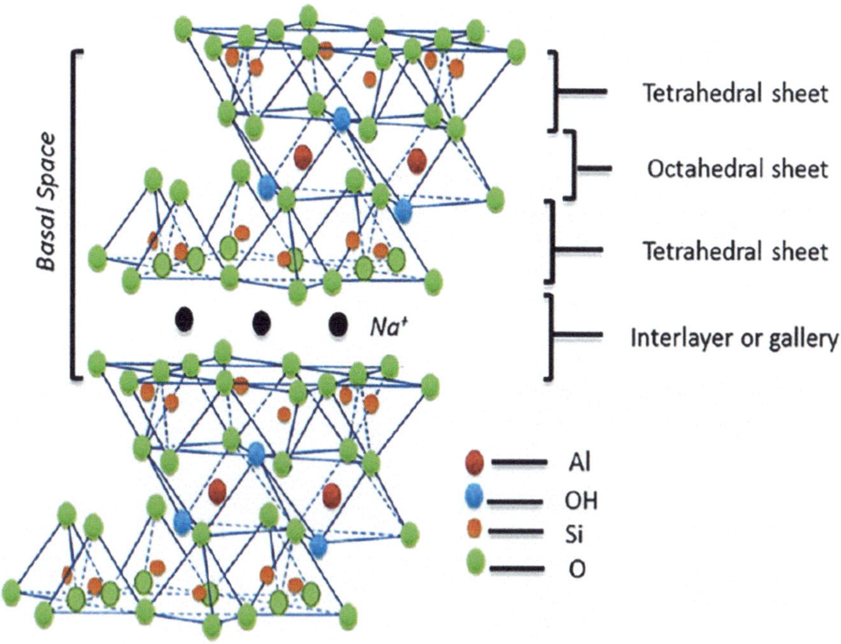

Fig. 1.2 Sheet-like layered structure of MMT clays [114]

Table 1.4 Chemical formulae and cation exchange capacity (CEC) of 2:1 phyllosilicates [106]

Silicate	Chemical formula	CEC (meq/100 g)
Montmorillonite	$M_x(Al_{4-x}Mg_x)Si_8O_{20}(OH)_4$	92.6–120
Hectorite	$M_x(Mg_{6-x}Li_x)Si_8O_{20}(OH)_4$	120
Saponite	$M_xMg_6(Si_{8-x}Al_x)Si_8O_{20}(OH)_4$	86.6
Vermiculite	$(Mg, Fe, Al)_3[(AlSi)_4O_{10}](OH)_2M_x{\cdot}nH_2O$	150

* M is the cation exchange and *x* is the degree of substitution

of organic molecules. In particular, MMT nanoclays are of wide use due to their easy modification with organic cations on the interlayer galleries. In addition, MMTs also have a good affinity with silicate and ammonium ion bonding [102, 116]. Nowadays, commercial organoclays are available in the market with a reasonable price. Table 1.5 listed current commercial products and manufacturers of organomodfied clays. Each of commercial organomodified clays has different modified cation, specific interlayer spacing and CEC leading to various product characteristics.

Table 1.5 Material types of commercial organoclays

Producer	Clay	Modifier agent	d_{001}(nm)	CEC (meq/100 g)	References
BYK Additives and Instruments (Germany)	Cloisite 10A	Dimethyl, benzyl, hydrogenate tallow, quaternary ammonium	1.92	125	[4, 41, 115]
	Cloisite 15A	Dimethyl, dihydrogenated tallow, quaternary ammonium	3.15	125	
	Cloisite 20A	Dimethyl, dihydrogenated tallow, quaternary ammonium	3.15	95	
	Cloisite 25A	Dimethyl, dihydrogenated tallow, 2-ethylhexyl quaternary ammonium	1.86	95	
	Cloisite 30B	Methyl, tallow, bis-2-hydroxyethyl, quaternary ammonium	1.85	90	
	Cloisite 93A	Methyl, dehydrogenated tallow ammonium	2.36	90	
	Cloisite Na^+	Sodium montmorillonite	1.17	90	
Nanocor Inc. (USA)	Nanomer I.28E	Octadecyl trimethyl ammonium	2.43	93.7	
	Nanomer I.30E	Octadecyl amine	1.82	145	
Laviosa Chimica Mineraria (Italia)	Dellite LVF	-	0.98	105	
	Dellite 43B	Dimethyl benzylhydrogenated tallow ammonium	1.86	95	
Co-op Chemicals, Japan	Somasif TM (ME-100)		1.25	120	
	MAE	Dimethyl dialkyl (tallow) ammonium	3.0	120	
	MTE	Trioctyl methyl ammonium	2.5		
	MEE	Methyl bis-2-hydroxyethyl coco quaternary ammonium	2.3	120	

(continued)

Table 1.5 (continued)

Producer	Clay	Modifier agent	d_{001}(nm)	CEC (meq/100 g)	References
	MPE	Polyoxy propylene methyl diethyl ammonium	0.5	–	
Sud-Chemie (German)	Optigel EX0255	Sodium montmorillonite	–	100	
	Nanofil 804	Stearyldiethoxyamine	1.8	21	

Note d_{001} denote the interlayer spacing of clay platelets in relation to the (001) reflection in XRD analysis

1.2.2.2 Halloysite Nanotubes (HNTs)

HNTs are different from MMT nanoclays with respect to their fundamental material structures. With the presence of an octahedral aluminium layer and a tetrahedral silicate layer having the hydrated characteristics, HNTs are also known as 1:1 phyllosilicate. Similar to a coil, the structure of HNTs is in form of multi-layered tubes rolled up by the layers of aluminosilicates, as observed in Fig. 1.3. HNTs belong to kaolin family with a chemical formula of $Al_2Si_2O_5(OH)_4{\cdot}nH_2O$ [117]. HNTs are naturally abundant and available as hydrated clays with $n = 2$ used for the water content. In general, the evaporation process of natural kaolin yields the HNT structure [118]. Lvov and Abdullayev [118] has also found that the HNT length is approximately 0.5–2 μm whilst the inner and outer diameters are about 10–30 and 50–70 nm, respectively. Moreover, eco-friendly HNTs have become more cost-competitive when compared with CNTs. Their unique cylindrical tubular structure

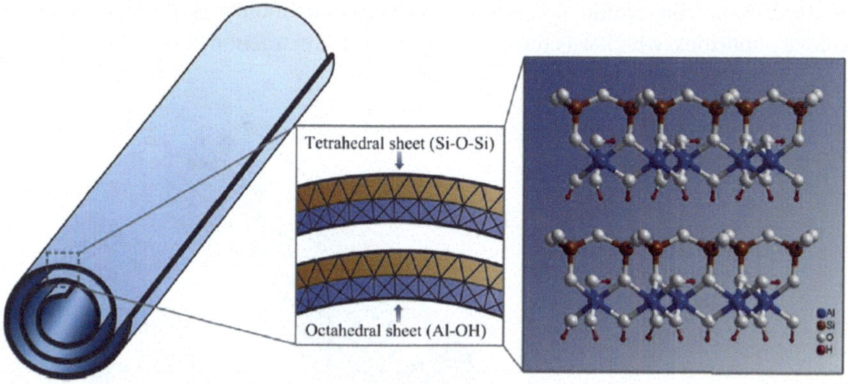

Fig. 1.3 HNT cylindrical tubular structures [120]

assists in the loading of chemical and biological substances, especially for drug delivery in medical applications [119].

HNTs are most widely used as nanofillers for polymer matrices within nanocomposites including poly(hydroxybutyrate-co-hydroxyvalerate) (PHBV) nanocomposites [121], epoxy nanocomposites [42, 107, 122–124], carboxylate butadiene–styrene rubber (xSBR) nanocomposites [108], polyamide 6 (PA6) nanocomposites [109], ethylene propylene diene monomer (EPDM) nanocomposites [110] and polyamide 12 nanocomposites [111]. Moreover, HNTs are also used as nanofillers for biopolymer matrices such as PLA [125], PLA/PCL [126], chitosan [127], polyethylene glycol/pectin [65], potato starch [128], soy protein [129], etc. In recent years, HNTs as clay nanofillers have also acted as a coating material for the anti-oxidation and corrosion of encapsulated inhibitors packed with a benzotriazole corrosion agent to protect the copper from highly corrosive media [130]. A range of metals such as iron, copper, cobalt, and nickel can form radical surfaces on HNTs for the production of electromagnetic materials [119]. Other applications include cosmetics, drug delivery [131] and bio-activeness [118].

1.2.2.3 Imogolite Nanotubes (INTs)

INTs are identical to HNTs in their tubular clay structure except for different particle sizes. INTs have a tubular length of typically 100 nm with inner and outer diameters of 1 and 2 nm, respectively. It has been stated that INTs offer a smaller structure when compared with HNTs with higher surface areas, which can be used to enhance the interaction with organic materials [132]. As demonstrated in Fig. 1.4, INTs consist of gibbsite-like sheets layered with silicate hydroxide and aluminium hydroxide on the inner and outer sides, respectively.

In 1962, natural imogolite clays were initially discovered in weathering pumice bed soils from volcanic ash [112]. Then it was followed with the purification process by eliminating iron, alumina, silica and carbonaceous moiety contents in order to remove impurities, which was further followed by the extraction of organic materials

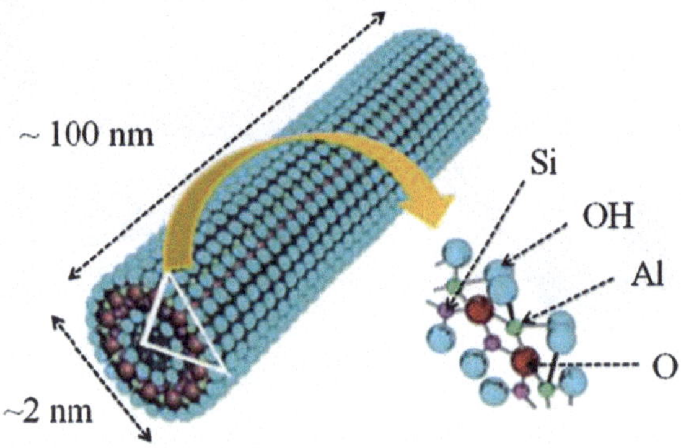

Fig. 1.4 INT tubular structures [132]

using acetic acid solutions, as studied by Imamura et al. [133] and Tsujimoto et al. [134]. Finally, imogolite gels were obtained by filtering the solutions containing imogolite prior to their dehydration calcination in air. The synthesis of INTs was introduced by mixing an aqueous solution of aluminium chloride ($AlCl_3 6H_2O$) and tetraethoxysilane ($Si(OEt)_4$) [135]. Since imogolite material can absorb more organic materials [136], Farmer's method becomes very popular for INT synthesis [137–141]. INTs possess hydrophilic properties with OH groups on the outer layer surfaces that can be easily dissolved in organic solvents and polymer matrices [142]. INTs can be employed as a stabiliser for platinum particles [138], biodegradation adsorbents [136, 143], catalysts [133, 144], which are considered as effective nanofillers of polymer nanocomposites [113, 140, 145–147].

1.2.3 Nanoclay Modifications

The interesting aspects of polymer/nanoclay nanocomposites lie in nanoclay wettability and modification, as well as interfacial adhesion between nanoclays and polymer matrices. The surfaces of organoclays can be tailored to alter their chemical properties and compatibility with polymer matrices. The cation exchange method is essential for surface modification by changing the cation ion with an organic surfactant, and further subjected to the modification from organophobicity to organophilicity. For example, Usuki et al. [26] modified MMT nanoclays by using cation exchange reaction with an amino acid solution. It was found that MMT interlayers were modified by an amino acid to produce new nanoclay intercalated structures. A typical case can be represented by the addition of modified MMTs on nylon 6 matrices for better nanoclay dispersion with the reduction of molecular weight (MW) of nylon 6/clay nanocomposites. A new network was further established between nylon 6 and MMT interlayer surfaces with the aid of ammonium cations.

Triantafillidis et al. [102] and Usuki et al. [26] used Jefamine diamine curing agent to modify nanoclay interlayers by ion exchange reaction. The increase of MW of an onium ion can affect the increase in clay basal spacing, as shown in Fig. 1.5. Jefamine onium ions with high MW and long structural chains can result in large clay basal spacing values. The basal spacing of deprotonated fluoro hectorite clays can be enlarged up to 4.6 and 1.38 nm when using Jefamine onium ions with a high MW of 2000 and a low MW of 230, respectively.

Nonetheless, Čapková et al. [149] tried to achieve Na^+ MMT intercalation with steric acid and amine. Meanwhile, Ha et al. [150] modified Na^+ MMT clays with 3-aminopropyltriethoxysilane solutions in three major steps including clay dispersion with solvents, mixing, as well as filtering and drying under vacuum. Kornmann et al. [148] evaluated the effect of different types of amino groups on protonated fluorohectorite (Somasif ME-100) silicate layers, which indicated that amino groups were successfully intercalated into silicate layers. It is also noted that MW of amino groups can affect the interlayer spacing of oganomodified clays.

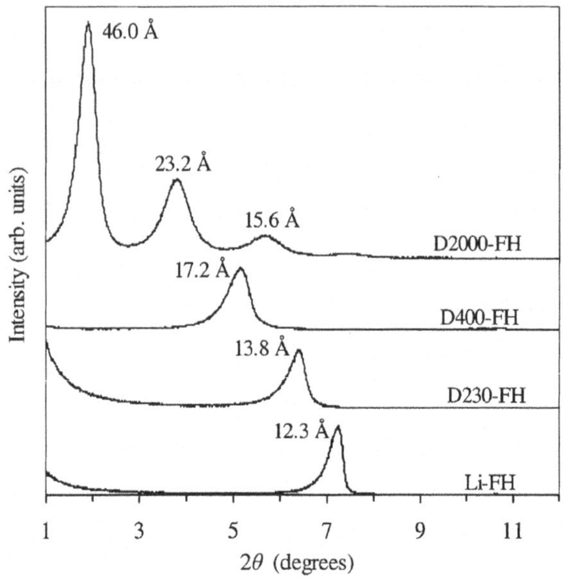

Fig. 1.5 XRD of fluoro hectorite clay deprotonated by a Jefamine onium ion [102]

On the other hand, the modification of clay interlayers or surfaces have also been conducted for HNTs and INTs. For instance, Li et al. [151] studied the modification of interlayer and outerlayers for natural HNTs with 3-aminopropyltrimethoxysilane (APS) using the atom transfer radial polymerisation (ATRP) method. Furthermore, Tang et al. [123] and Tang et al. [124] utilised phenylphosphonic acid (PPA) to modify unfolded HNTs. The modifications on HNT layers were found to improve HNT wettability in epoxy matrices. The HNT clays were dispersed well with a combination of partially and fully intercalated structures. Similar results were found by Yamamoto et al. [113] and Ma et al. [145] when INTs were modified with methyl methacrylate. Furthermore, Park et al. [152] also conducted INT modification by using octadecyl phosphonic acid.

1.2.4 Nanoclay Dispersion Status

Morphological structures of polymer/clay bionanocomposites rely on processing conditions and methodologies of material synthesis. Alexandre and Dubois [10] reviewed the structures of polymer/clay nanocomposites and indicated three categories of immiscible structure to form conventional microcomposites, intercalated and exfoliated structures, as depicted in Fig. 1.6.

When nanoclay aggregates are separated from polymer matrices without the change of clay basal spacing, this results in microcomposites due to micron-sized clay aggregates. As polymeric molecules are inserted into nanoclay particles to broaden clay interlayer areas, intercalated structures take place. While polymer molecules

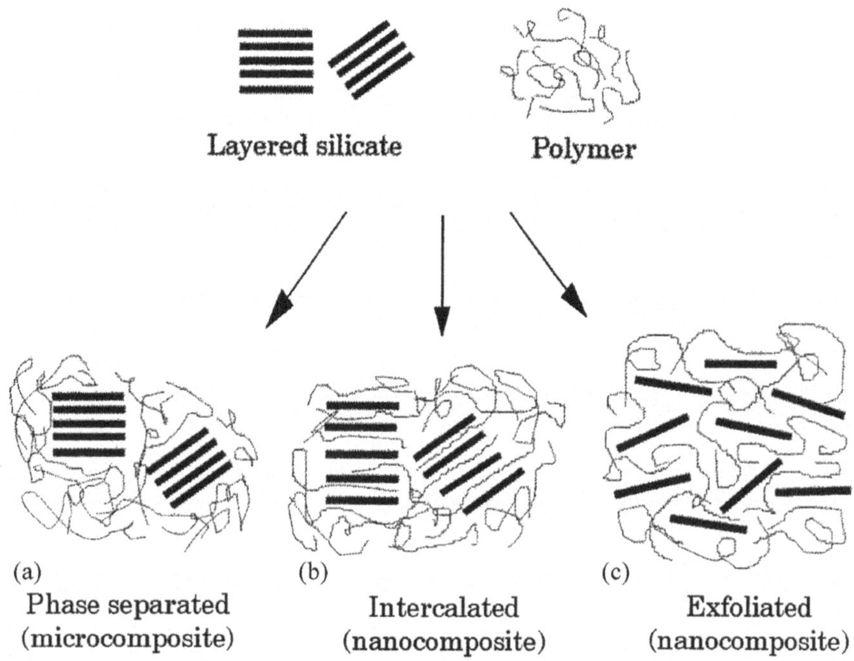

Fig. 1.6 Different types of clay structures when dispersed into polymer matrices, **a** conventional microcomposite structure, **b** intercalated structure, **c** exfoliated structure [10]

are diffused into nanoclay galleries with sufficient interlayer spacing to allow for the dispersion of individual clay layers, exfoliated structures are generated with the superior properties of nanocomposites.

Nanoclay dispersion degree or status within polymer matrices would potentially influence the improvement of mechanical, thermal and physical properties of nanocomposites. The dispersion status is generally studied by XRD analysis such as either wide-angle X-ray diffraction (WAXRD) or small-angle X-ray scattering (SAXS) and transmission electron microscopy (TEM) [153, 154]. By performing XRD analysis, the peak characteristics of clay platelet interlayers can be observed at low diffraction angle. Meanwhile, TEM images offer an actual observation for the clay dispersion such as clay aggregation, exfoliated or intercalated structures in terms of clay interlayer spacing and orientation degree of clay layered structures.

Current studies indicate that MMTs and HNTs can also be incorporated into plant oil based biopolymers. Miyagawa et al. [90] reported the compatibility of bio-based epoxy (i.e. ESO and ELO with organomodified MMT (OMMT) clays Cloisite 30B. The addition of 5 wt% Cloisite 30B clays could achieve completely exfoliated structures and homogenous clay dispersion (Fig. 1.7). In addition, Liu et al. [81] and Zhu and Wool [155] indicated that a mix of intercalated and exfoliated structures occurred with different nanoclay contents and types. On the other hand, Tan

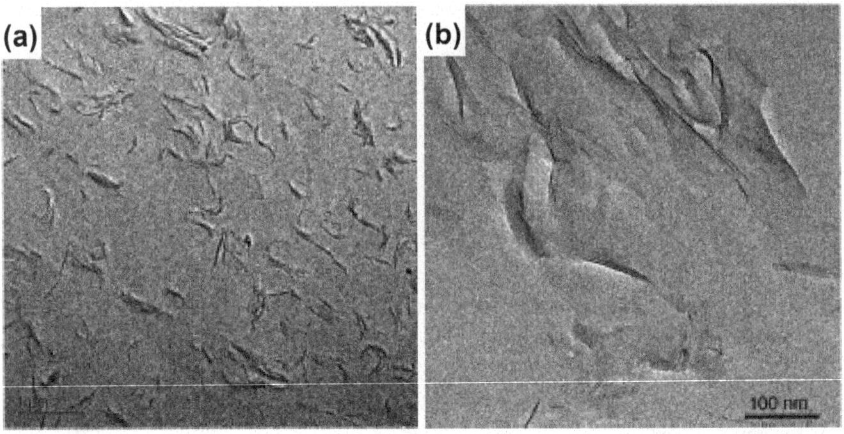

Fig. 1.7 TEM micrographs of bioepoxy/clay nanocomposites at **a** low magnification (scale bar: 1 μm), **b** high magnification (scale bar: 100 nm) [90]

et al. [156] reported physicochemical properties of organomodified HNT/epoxidised natural rubber (ENR) nanocomposites.

In addition, Wang et al. [22] observed the formation of nanostructures by using SAXS and TEM analyses. The SAXS results indicated that well intercalated (with the possible exfoliation) or highly intercalated structures were observed with the addition of over 2 wt% OMMT clays. Meanwhile, at high clay loadings of 4 and 6 wt%, clay peak broadening effect became quite manifested. On the other hand, TEM analysis of those nanocomposites revealed the existence of combined disordered exfoliated structures and microsized clay aggregates.

1.3 Material Fabrication and Optimisation of Biopolymer/Clay Nanocomposites

1.3.1 Solution Intercalation

Solution intercalation or solution mixing is based on the solubility of bio-based epoxy matrices and appropriate solvents. Commonly used solvents include acetone, chloroform, methanol and ethanol, as listed in Table 1.6. The compatibility between solvents, nanoclays and polymer matrices depends on several factors such as solvent polarity, evaporation temperature, solvent volatility and dispersion degree with high amounts of nanoclays [34, 157].

During the fabrication process, clay nanofillers are initially swollen in the solvent prior to being mixed with bioepoxy matrices. Subsequently, bioepoxy molecules are intercalated into clay interlayers to substitute for solvent molecules that are eventually

Table 1.6 Summary of different solvents used in solution intercalation [132]

Resin	Solvent	Clay	Description	References
Thermoplastic polyurethan (TPU)	Dimethylacetamide (DMAc) Tetrahydrofuran (THF)	Cloisite 30B Cloisite 15A Cloisite 25A	• DMAc has better compatibility with Cloisite 30B • THF has better compatibility with Cloisite 15A • DMAc and THF have a similar approach on Cloisite 25A	[164]
Diglycidyl ether of bisphenol A (DGEBA)	Acetone	S30A, Southern Clay Products,Inc B34,Rheox, Inc B24,Rheox, Inc	• Enhance mixing process and dispersion process • Solvents do not influence the structures and final properties of nanocomposites	[165]
DGEBF/ functionalised vegetable oils (FVOs)	Acetone	Cloisite 30B	• Homogenous dispersion of clay fillers	[90]
Diglycidyl ether of bisphenol F (DGEBF)/ epoxidised linseed oil (ELO)	Acetone	Cloisite 30B	• Clays are dispersed well into biopolymer matrices	[51]
Conjugated corn oil (CCOR) styrene (ST)/divinylbenzene (DVB)	Triethyl(4-vinylbenzyl)ammonium chloride (VTAC)	Na-Montmorillonite (MMT)	• VTAC can enlarge clay interlayer spacing	[163]
Unsaturated polyester/epoxidised methyl soyate	Acetone	Cloisite 30B	• Excellent dispersion and exfoliation of organoclay particles • Waste of time, cost and energy • Influence decreased tensile properties	[34]
PLA	Chloroform	Cloisite 30B	• Supporting PLA to penetrate into clay layer	[54]
1,4:3,6-Dianhydro-D-sorbitol (isosorbide)-based polycarbonate	DMAc	Na-MMT, Kunipia-F Cloisite 30B	• Homogeneous organoclay dispersion in polymer matrices	[166]

(continued)

Table 1.6 (continued)

Resin	Solvent	Clay	Description	References
Poly(3-methacrylamidopropy(N,N-dimethyl) dodecyl ammonium bromide)	Ethanol Chloroform Propylene glycol monomethyl ether (MeOPrOH)	Cloisite Na (ClNa) Cloisite 15A (Cl15A) Cloisite 20A (Cl20A) Cloisite 30B (Cl30B) Cloisite 93A (Cl93A)	• Solvents influence the appearance of final products • Ethanol solvent results in a heterogeneous polymer–clay system • Chloroform and MeOPrOH result in gel aspect and clear solution • Gel volume for chloroform:Cl15A > Cl20A > Cl93A > Cl30B > ClNa • Gel volume for MeOPrOH: Cl30B > Cl20A > Cl15A > Cl93A > ClNa	[157]

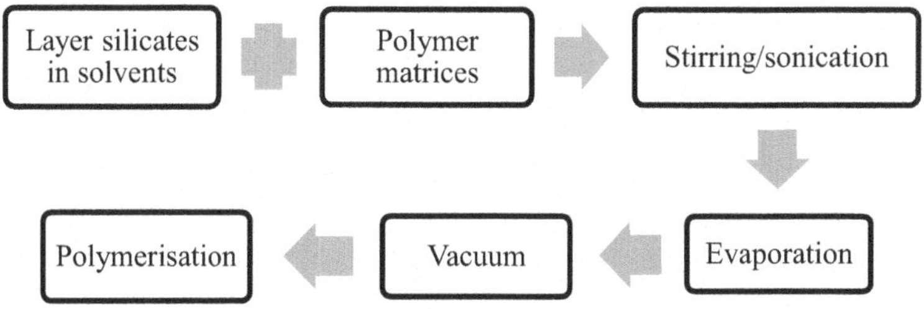

Fig. 1.8 Flowchart process of solution intercalation method [132]

evaporated. Upon the complete solvent removal, intercalated structures are formed in the bionanocomposite system with the aid of a continuous polymerisation process, Fig. 1.8. The main reason behind the solution intercalation method is to increase the nanoclay dispersion, thus resulting in the improvements of physical, mechanical and thermal properties, as presented by Alexandre and Dubois [10] and Lebaron et al. [158].

Considerable studies on nanocomposite materials using the solution intercalation method have been reported by previous researchers for conventional polymer/clay nanocomposites [90, 159–162]. Hutchinson et al. [160] compared two different methods, direct mixing and solution intercalation of epoxy/clay nanocomposites by using acetone as a solvent. According to micrographic results conducted by SEM analysis, the solvent method achieved much better dispersion compared to direct mixing. The largest agglomerates were only about 20 μm in size, as opposed to 80 μm by direct mixing. In addition, the studies concerning the synthesis of biodegradable polymer/clay nanocomposites by the solution intercalation method have also been reported in Tables 1.7 and 1.8. Lu and Larock [163] investigated the effect of functionalised triethyl(4-vinylbenzyl)ammonium chloride (VTAC) surfactant on the dispersion of Na^+ MMT nanoclays in conjugated corn oil (CCOR) biopolymer. The functionalisation of VTAC surfactant can increase the clay interlayer spacing from 1.15 to 1.51 nm. On the other hand, Haq et al. [34] worked with styrene-based unsaturated polyester, epoxidised methyl soyate, epoxidised soybean oil and platelet-like nanoclays by solution mixing where acetone acted as a solvent. The use of acetone in the fabrication of bio-based polymer/clay nanocomposites resulted in good nanoclay dispersion and improved tensile modulus of such nanocomposites. The functionalised acetone can facilitate organoclay dispersion in polymer matrices and produce a homogeneous mix of materials.

1.3.2 In Situ Polymerisation

In in situ polymerisation process, clay nanofillers are directly mixed with polymer matrices. The monomers are then diffused into clay interlayers to produce clay intercalation structures. The final polymerisation process can be achieved either by heat

Table 1.7 Summary of current study in the fabrication of thermosetting biopolymer nanocomposites

Polymer	Curing agent/catalyst/ plasticiser/solvent	Filler type	Process	Structure status	References
ESO	TETA	Cloisite 30B	In situ polymerisation	Intercalated	[32]
Acrylated epoxidised soybean oil (AESO)	Maleic anhydride/1.5 wt % tert-butyl peroxy benzoate	Cloisite 30B	In situ polymerisation	Exfoliated	[167]
Maleinised acrylated epoxidised soybean oil (MAESO)		Cloisite 30B		Intercalated	[167]
Soybean oil pentaerythritol maleates (SOPERMA)		Cloisite 30B		Intercalated	[167]
ESO	MTHPA/1-methylimidazole	OTAC	In situ polymerisation	Intercalated	[27]
		Bengel 434	In situ polymerisation	Intercalated	[27]
	Benzylsulfonium hexaflouroantimonate derivative	OMMT	In situ polymerisation	Intercalated/exfoliated	[168]
ELO	Benzylsulfonium hexaflouroantimonate derivative	OMMT	In situ polymerisation	Intercalated/exfoliated	[168]
Acrylated epoxidised castrol oil (AECO)/DGEBA	TETA	Cloisite 30B	Solution intercalation	-	[168]
ESO/DGEBF	MTHPA/1-methylimidazole	Cloisite 30B	Solution intercalation	Exfoliated	[90]
ELO/DGEBF		Cloisite 30B	Solution intercalation	Exfoliated	[90]

(continued)

Table 1.7 (continued)

Polymer	Curing agent/catalyst/ plasticiser/solvent	Filler type	Process	Structure status	References
AESO/styrene (50/50)	2,20-Azoisobutyronitrile (AIBN)	Nanofil 1080	In situ polymerisation	Exfoliated	[31, 170]
Low-saturation (LSS) soybean oil/styrene (ST)/divinyl- benzene (DVB) (50/25/10)	Boron trifluoride diethyl etherate (BFE) (15)	Modified-MMT	Cationic polymerisation	Intercalated	[171]
Diglycidylether bisphenol-S (DGEBS)/Hyperbranched polyurethane (HBPU) (100/30)	poly(amido amine)	OMMT	In situ polymerisation	Intercalated/exfoliated	[172]
M. ferrea L. seed oil/glycerol/Epichlorohydrin/tetra bromobisphenol-A (1/16/2/1)	poly(amido amine)	Nanomer I.30E	Ex situ polymerisation	Exfoliated	[173]
Epoxidised M. ferrea L. seed oil/Bisphenol-A (50/50)	poly(amido amine)	OMMT	Ex situ polymerisation	Exfoliated	[54]
PBS/DGEBA	1,4-butanediol/succinic acid	Twice functionalised organoclay (TFC)	Solvent intercalation	Intercalated	[174]
Glycerol polyglycidyl ether (GPE)/polyglycerol polyglycidyl ether (PGPE)	ε- poly(L-lysine) (PL)	Kunipia F	In situ polymerisation	Intercalated/exfoliated	[28]

Table 1.8 Summary of biopolymer nanocomposites based on thermoplastic polymer matrices

Polymer	Filler type	Process	Structure status	References
Poly(ester amide) (PEA)	Nanomer 1.30E. (MMT)	Melt intercalation	Intercalated	[175, 176]
	Cloisite 25A	Melt intercalation	Intercalated	[177]
	Nanofil 848	In situ polymerisation	Intercalated	[178]
	Cloisite 20A, Cloisite 25A, Cloisite 30B	In situ polymerisation	Intercalated/exfoliated	[178]
poly (butylene succinate-co-adipate) (PBSA)	HNT	Melt intercalation	Dispersed finely	[179]
	Cloisite 15A	Melt intercalation	Exfoliated	[179]
	Twice-functionalised organoclay (TFC)	Melt intercalation	Intercalated	[180]
Starch	Cloisite Na^+	Solution intercalation	Intercalated/partially exfoliated	[181]
poly(ε-caprolactone) (PCL)	Cloisite 25A	Melt intercalation	Intercalated	[182, 183]
	Cloisite 30B	Melt intercalation	Exfoliated	[184–186]
	Cloisite 20A	Melt intercalation	Exfoliated	[187, 188]
	Na^+MMT	Solvent intercalation	Exfoliated	[189]
	Imogolite	Solvent intercalation	–	[190]
poly (butylene succinate) (PBS)	MMT	Melt intercalation	Intercalated	[191]

(continued)

Table 1.8 (continued)

Polymer	Filler type	Process	Structure status	References
	O-MMT	Melt intercalation	Intercalated	[191]
	MMT-cetyl pyridinium chloride	–	Intercalated	[192]
	MMT-cetyl trimethyl ammonium bromide	–	Intercalated	[192]
	Cloisite 15A	Solution and melt intercalation	Intercalated	[193]
	Cloisite 15A	Melt state using a torque rheometer (SOAM)	Intercalated	[193]
	HNT	Melt intercalation	Uniform distribution	[194]
PLA	HNT	Melt intercalation	Aggregated at > 5wt%	[195]
	Cloisite 30B	Solution intercalation	Exfoliated	[196, 197]
	Cloisite 10A	Solution intercalation	Exfoliated	[198]
	HNT	Solution casting	Better dispersion	[199]
PLA/epoxidised natural rubber	HNT	Melt intercalation	nanodispersion/agglomerated	[200–202]
PLA/poly(butylene adipate-co-terephthalate (PBAT)	Nanomer I.34	Melt intercalation	Intercalated/exfoliated	[203]
PLA/PBSA	Cloisite Na^+	Melt intercalation	Intercalated	[204]
	Cloisite 15A	Melt intercalation	Disordered structure, Intercalated/exfoliated	[204]
	Cloisite 20A	Melt intercalation	Intercalated	[204–206]
	Cloisite 25A	Melt intercalation	Disordered structure, Intercalated/exfoliated	[204]
	Cloisite 30B	Melt intercalation	Disordered structure, Intercalated/exfoliated	[204]
PLA/poly(methyl methacrylate) (PMMA)/epoxidised natural rubber	HNT	Melt intercalation	Uniform dispersion	[207]

(continued)

Table 1.8 (continued)

Polymer	Filler type	Process	Structure status	References
PLA/TPU	HNT	Melt intercalation	–	[208, 209]
PLA/starch	MMT	Premixing/extruder	Intercalated/exfoliated	[210]
PLA/PCL	HNT	Solution intercalation	Intercalated	[211]
PLA/PBSA	Cloisite 25A	Melt intercalation	Intercalated/exfoliated	[180]
PLA/PBS	Twice functionalised organoclay (TFC)	Melt intercalation	Intercalated/exfoliated	[180]
	Cloisite 30B	Melt intercalation	Intercalated/exfoliated (< 3wt%), agglomerated (> 3wt%)	[212]

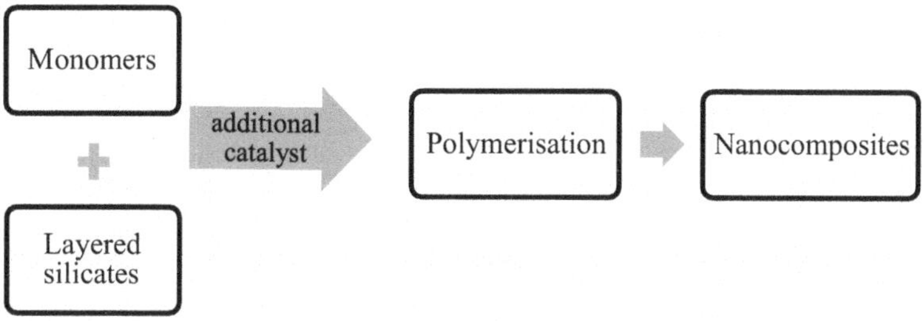

Fig. 1.9 Flowchart of in situ polymerisation method [132]

or radiation or by the diffusion of initiators or catalysts [3, 4]. The flowchart of in situ polymerisation is illustrated in Fig. 1.9. The processing conditions are controlled by temperature, mixing speed, mixing time and shear stress generated from the rotor or blade of a mixer [10]. The merit of this process lies in the adaption to current industrial polymer processing without using unrecyclable and harmful organic solvents.

The current studies in the fabrication of biopolymer/clay nanocomposites by in situ polymerisation were summarised in Tables 1.7–1.9. By applying in situ polymerisation method, it is found that the solubility of clay nanofillers has been improved in biopolymer matrices. Uyama et al. [168] reported that the combination of intercalated and exfoliated structures was evident in ESO/OMMT bionanocomposites produced by in situ polymerisation. In addition, Liu et al. [32] also found that good clay dispersion could be achieved without any clay aggregates in ESO matrices using in situ polymerisation.

1.3.3 Melt Intercalation

Melt intercalation is based on the melting point of polymer matrices and applied by annealing above the softening point of polymers [4]. In the annealing stage, bulk polymers melt and then diffuse into the galleries between nanolayers and nanoparticles. The fabrication steps of melt intercalation technique for polymer nanocomposites can be seen in Fig. 1.10. Melt intercalation technique is widely used in nanocomposites fabrication due to its compatibility with current industrial processes such as injection moulding and extrusion processes. Moreover, these techniques also offer some typical benefits, which is more ecofriendly due to the generation of solvent-free wastes.

The study of melt intercalation and final nanocomposite structures are much more influenced by the compatibility of coupling agents, as well as processing conditions. Zhao et al. [213] has reported that bionanocomposite structures can be intercalated or exfoliated, which is influenced by the interaction between polymer matrices and fillers. The goal of melt intercalation is to achieve high clay wettability in polymer

Table 1.9 Methods and properties of bionanocomposite materials using in situ polymerisation [132]

Materials	Method	Description	Properties	References
ESO/octadecyl-modified montmorillonite (OMM)	Thermal latent catalyst + heat	• During thermal treatment, crosslinking occurred	• The addition of 5 and 15 wt% OMM showed intercalated and randomly distributed structures in polymer matrices • Increase T_g with increasing the clay content from 5–15 wt%, ranging from -2 to 4°C	[168]
ESO/MMT	Initiator + heat	• The ratio of epoxy and initiator influenced the crosslinking density and material properties	• Good dispersion without organoclay aggregation • High storage modulus, tensile strength and tensile modulus were reached at 8 wt% organoclay inclusions • The best ration of epoxy: amine = 1:1.37 • Thermal stability was above 180°C	[32]
Acrylated oleic methyl ester (AOME) bio-based elastomer/MMT + Methyl methacrylate (MMA)/ ethylene glycol dimethacrylate (EGDMA)	Cationic catalyst	• Curing at room temperature by cationic catalyst cobalt naphthenate and cumyl hydroperoxide in solution • MMA and EGDMA as crosslinking agents	• Solubility parameter was influenced by hydroxyl groups in modified clay and carbonyls in poly(AOME) • Cloisite 30B and poly(AOME) achieved the highest degree of intercalation • Tensile strength and elongation at break increased with additional clays from 1 to 10 wt% • Additional MMA increased tensile strength, while strain to failure decreased • T_g of the sample with 10 wt% clays varied from -56°C to -48°C • Thermal stability increased with increasing the clay content • Soil burial test showed a biodegradable behaviour of AOME/clay nanocomposites	[155]

(continued)

Table 1.9 (continued)

Materials	Method	Description	Properties	References
Conjugated corn oil (CCOR)/ Triethyl(4-vinylbenzyl)ammonium chloride (VTAC)-MMT	Cationic catalyst and initiator	• Styrene and divinylbenzene as crosslinking agents	• Modified MMT clays with VTAC improved the clay dispersiblity level in oil compared to natural MMTs • High viscosity and flocculation limit of VTAC-MMT fillers influenced the dispersibility of fillers in reaction • Compressive strength reached the maximum value with addition of 2 wt% fillers • Thermal stability increased with increasing the clay loading • Crosslinking agent enhanced the strength with decreasing the strain to failure	[163]
Mesua ferrea L. seed oil/bisphenol-A/ tetrabromobisphenol-A/MMT	Initiator	• Clay content influence to curing condition	• Morphology with uniform and stable clay dispersion • Viscosity inevitably became higher with increasing the clay loading • Improvements in performance characteristics like curing, mechanical, thermal and chemical resistance was achieved with increasing clay loading	[54]

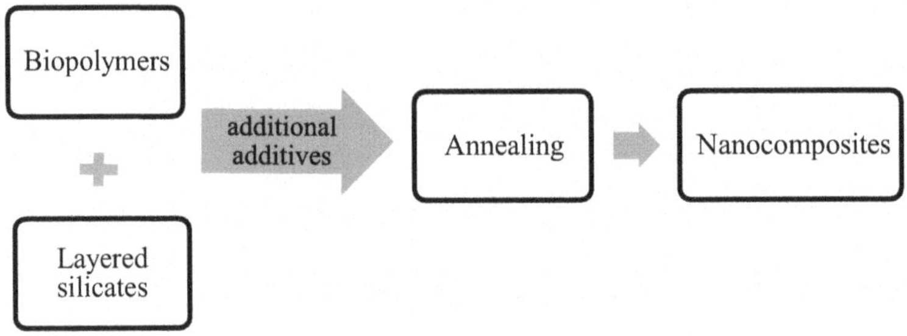

Fig. 1.10 Flowchart of melt intercalation method

matrices, which is indicative of homogeneously dispersed fillers into matrices leading to fully exfoliated or homogenous intercalated structures.

On the other hand, Vaia and Giannelis [218] investigated the theoretical study on achieving optimal fabrication of nanocomposites by melt intercalation. It is noted that interlayer chemistry, molecular weight and annealing temperature are very important in the formation of clay exfoliated structures. Moreover, polymeric molecular motion can lead to their diffusion into clay interlayers. Current research on manufacturing polymer nanocomposites using melt intercalation method can be seen in Table 1.10. Huang et al. [214] employed melt intercalation method to fabricate poly(etherimide) (PEI)/nanocomposites. It was found that fully exfoliated structures occurred, as confirmed by XRD analysis and TEM images. However, other studies revealed the combination of intercalated and exfoliated structure of organoclays in different types of polymer matrices using melt intercalation [183, 215, 216]. Hence, the compatibility of polymer matrices and clay fillers can affect the former nanocomposite structures.

1.4 Optimisation Technique and Effective Synthesis of Bio-based Polymer/Clay Nanocomposites

The properties of polymer/clay bionanocomposites can be affected by the compatibility and reactivity of biopolymers and nanoclays, their manufacturing methods and processing parameters [22, 28, 32–36]. Several processing parameters, influencing the interactions between nanoclay particles and polymeric molecules, are known as follows: (i) nanoclay dispersion [44–47], (ii) nanoclay modification [102, 148, 150, 219], (iii) compatibility between nanoclays and polymer matrices [39–43], and (iv) curing agents for thermosetting materials [11, 37, 38]. Consequently, a major concern is how to understand the modification mechanism of natural nanoclays.

The mechanism of clay dispersion in polymer matrices can be affected by mixing types used in the pre-mixing stage. Among pre-mixing processing techniques between fillers and polymer matrices, mechanical shear mixing and sonication have been employed in effectively dispersing clay fillers into polymer matrices [16, 22,

Table 1.10 Summary of current research in bionanocomposites using melt intercalation

Polymer	Filler	Additive	Description/properties	References
Poly(etherimide) (PEI)	Na-MMT modified with 1-Hexadecylamine	–	• Homogenously exfoliated structures were proved by XRD and TEM analyses • Thermal decomposition temperature (T_d) and T_g of PEI/clay nanocomposites were increased with increasing the clay loading • The incorporation of organoclays reduced solvent uptake properties	[214]
poly(ε-caprolactone) (PCL)	Na-MMT hydrogenated tallowalkyl (HTA)-based quaternary ammonium cations	–	• A processing technique by using a conventional two-roll mill at 130 °C • A combination of intercalated/exfoliated structures observed by XRD and TEM analyses • Stiffness increased up to 5 wt% in clay content and decreased with further increasing the clay loading	[183]

(continued)

Table 1.10 (continued)

Polymer	Filler	Additive	Description/properties	References
PLA	MMT	–	• The dispersion of nanoclays was quantified by the solid state of nuclear magnetic resonance (NMR) • Optimal processing condition was achieved by applying the short time of mechanical energy, followed by low shear stress for 5 min and final delamination of clay platelets	[33]
Thermoplastic polyurethane (TPU)	OMMT	–	• Melt mixing is an effective method for dispersing OMMT throughout the TPU matrix • Homogenous clay dispersion with combination intercalated and exfoliated structures • T_g increased with increasing the clay loading • The addition of 5 wt% MMT clays was found to improve ultimate tensile strength, tensile modulus and elongation at break by approximately 150, 230 and 25%, respectively	[215]

(continued)

Table 1.10 (continued)

Polymer	Filler	Additive	Description/properties	References
High density polyethylene (HDPE)	Modified bentonite clays	Polar compatibiliser Polybond 3009 (PE-g-MA)	• Counter-rotating and co-rotating twin-screw extruder types were used • Well-dispersed intercalated structure and partially exfoliated structure were discovered • The presence of organoclays, compatibiliser and the rotating extruder types were not sufficient to achieve ful clay exfoliation	[216]
Polypropylene (PP)	NANOLIN™ DK4 organomodified clays	Maleated polypropylene (MAPP)	• MAPP was employed as an effective compatibiliser to improve clay dispersion • A co-rotating intermesh twin screw extruder was used with a screw speed of 200 rpm at the temperature profile of 185–210 °C	[43]
Polyamide	MMT	–	• Twin screw microextruder applied for 2 min at 60 rpm, followed by blending for 6 min at 180 rpm at a melt temperature of 245 ± 3 °C • Extruded at 80 rpm into a heated transportable chamber • Injected at 0.8 MPa into a mould cavity • α structure of polyamide and exfoliated structures were identified	[217]

28, 35 51–53]. In mechanical shear mixing, clay aggregates can be broken up into finer clay particles by impellers and fluid shear forces. Atiemo-Obeng et al. [55] revealed that operation methods, solid-phase and liquid-phase states and properties, unit operation, impeller geometry and mixing parameters are key factors in a solid–liquid mixing process. Alexandre and Dubois [10] also reproted that processing parameters in an effective control including mixing speed, mixing temperature and time and impeller shear forces became significant for the effective manufacturing of polymer/clay nanocomposites. However, Gupta et al. [220] applied a combined mechanical and high shear mixing process to disperse MMT clays into epoxy matrices, which resulted in exfoliated clay structures in nanocomposite systems. Subsequently, the functionalisation of shear mixing can also be used as a primary processing step for effective clay dispersion [220, 221].

On the other hand, the sonication method can be utilised for the further step to disperse fine clay particles after initial mechanical shear mixing. Air bubbles, resulting from the propagation of sound waves under ultrasonication, help polymer solutions to penetrate into clay interlayer galleries, thus leading to an increase in clay interlayer spacing [47]. In line with the solid–liquid dispersion techniques, ultrasonic parameters such as ultrasonic frequency, power intensity, temperature, vapour and external pressure, viscosity and surface tension of solid–liquid mixtures have great impacts on clay wettability and dispersibility [222]. Haq et al. [34] found that different parameters and techniques such as solvent type, the sequence of bio-resin addition, sonication energy and processing time influenced final tensile properties and morphological structures of bio-based unsaturated polyester/clay nanocomposites. In addition, Kaboorani et al. [47] indicated that increasing the power intensity of sonication up to 100% could improve clay wettability in less than 20 min for sonication time. Additionally, it was claimed that clay wettability and exfoliated structures could not be obtained solely based on ultrasonication. Hence, the combination of shear mixing and ultrasonication has been proven to be a key-processing step for the improvement of clay wettability according to Dean et al. [223].

On the other hand, mechanical milling technique can be potentially employed to improve clay dispersion in polymer matrices [224]. Lonkar et al. [225] noted that ball milling process did not require the use of solvents or high temperatures. Koo et al. [226] investigated the structure of polyvinylpyrrolidone (PVP)/Na^+-MMT nanocomposites via attrition ball milling. Exfoliated structures of clay nanocomposites could be achieved with the additional clays up to 20 wt%, beyond which intercalated structures were observed. Lu et al. [167] compared two different types of dispersion techniques, namely high-speed emulsification and homogenising mixer (HEHM) and ball milling. DGEBA resin and organomodified MMT clays were used for the material fabrication of nanocomposites while XRD and TEM were employed to analyse nanocomposite structures. Partial clay exfoliation and clay aggregation were observed for nanocomposites at the clay content of 3 wt% using a HEHM technique, resulting in lower impact and flexural strengths when compared with those of neat epoxy resin. Meanwhile, at a similar clay content, such nanocomposites prepared by the combination of HEHM and ball mill tended to be fully exfoliated with higher flexural and impact strengths than those of neat DGEBA resin. In addition, Ramadan et al. [227] investigated the effect of ball milling process on Cloisite Na^+ and Cloisite 30B clay structures. High-energy ball mill could easily delaminate platelet structures of Cloisite Na^+ clays to form exfoliation structures. Meanwhile, as for those clays with organomodified cations, Cloisite 30B clays did not show the peel-off effect with respect to platelet-like clay layered structures after 3 h ball milling process at 400 rpm. Additionally, other studies on the fabrication of nanocomposites by a ball mill process were also performed on polyimide-6/natural clay nanocomposites with pan milling equipment [208], PP/clay nanocomposites [209], epoxy/MMT nanocomposites [230] and epoxy/unsaturated polyester (UP) nanocomposites reinforced with organomodified clays [231].

1.5 Nanocomposite Properties

1.5.1 Mechanical Properties

Mechanical properties of polymer/clay nanocomposites were initially evaluated by Toyota R & D reserachers in 1990s [26, 232]. This study motivated many reserchers to work in the field of advanced polymer/clay nanocomposites. In particular, the investigation of the effect of nanoclay fillers on tensile strength [45, 150, 233], Young's modulus [37, 148], flexural properties of petroleum-based polymer/clay nanocomposites [40]. Moreover, numerous reviews on the functionalisation of nanofillers, the variation of matrices and additives, as well as the compatibility of coupling agents to affect the properties and processing conditions of fabricated nanocomposites have been mentioned elsewhere [1, 10, 39, 52, 106, 158, 234–241].

On the other hand, the addition of clay nanofillers into biopolymer matrices has also influenced the improvement of mechanical properties of biopolymer/clay nanocomposites. Mechanical properties of biopolymers filled with clays are listed in Tables 1.11 and 1.12. Uyama et al. [168] synthesised green nanocomposites using plant oil based biopolymers and MMT nanoclays, which were cured by latent acid catalyst benzylsulfonium hexafluoroantimonate. The tensile modulus and tensile strength of nanocomposites were significantly increased by 200% with 8 wt% nanoclay inclusions. Meanwhile, Miyagawa et al. [90] studied the effect of OMMT nanoclays on elastic modulus, impact strength and fracture properties of ELO/epoxy nanocomposites and ESO/epoxy nanocomposites. The exfoliated structures with 5 wt% organoclays improved the storage modulus by approximately 30%. However, intercalated nanocomposites with 5 wt % organoclays yielded 8 and 29% increases in fracture toughness when compared to those exfoliated nanocomposites and neat epoxy, respectively. Furthermore, the impact strength of nanocomposites decreased by about 16% with the addition of epoxidiser and organclays as opposed to that of neat epoxy.

Lu et al. [167] used new functionalised triglycerides plant oil to be combined with styrene as polymer matrices for bionanocomposite synthesis. The results showed

Table 1.11 Mechanical and thermal properties of biopolymer/clay nanocomposites

Polymer	Curing agent/catalyst/plasticiser/solvent	Filler type and content (wt%)	Mechanical properties				Thermal poperties			Reference
			Tensile strength (MPa)	Tensile modulus (MPa)	Elongation at break (%)	Impact strength (J/m)	T_g (°C)	T_m (°C)	T_d (°C)	
PLA	Dichloromethane (DCM)/dimethylformamide (DMF)	HNT (0)	–	–	–	–	59.28	153.77	–	[125]
		HNT (0.5)	–	–	–	–	60.2	169.38	–	
		HNT (1)	–	–	–	–	59.35	169.54	–	
		HNT (3)	–	–	–	–	60.1	170.98	–	
		HNT (5)	–	–	–	–	59.54	154.11	–	
	-	Cloisite 30B (0)	34	810	–	–	47.4	–	349.3	[197]
		Cloisite 30B (1)	38	940	–	–	53.2	–	376.7	
		Cloisite 30B (2)	–	–	–	–	55.1	–	386.3	
		Cloisite 30B (3)	–	–	–	–	51.3	–	393.8	
		Cloisite 30B (4)	–	–	–	–	49.9	–	–	
		Cloisite 30B (5)	–	–	–	–	48.9	–	–	
PCL	-	Cloisite Na^+ (0)	37	216	746	–	–	–	–	[183]
		Cloisite Na^+ (1)	35	201	715	33	–	–	–	
		Cloisite Na^+ (3)	35	197	714.5	22	–	–	–	

(continued)

Table 1.11 (continued)

Polymer	Curing agent/catalyst/plasticiser/solvent	Filler type and content (wt%)	Mechanical properties				Thermal poperties			Reference
			Tensile strength (MPa)	Tensile modulus (MPa)	Elongation at break (%)	Impact strength (J/m)	T_g (°C)	T_m (°C)	T_d (°C)	
		Cloisite Na+ (5)	32.5	207	666	19	–	–	–	
		Cloisite Na+ (10)	28	238	569	15	–	–	–	
	-	Cloisite 30B (1)	36	259	705	33	–	–	–	[183]
		Cloisite 30B (3)	35	272	563	18	–	–	–	
		Cloisite 30B (5)	34	313	560	13	–	–	–	
		Cloisite 30B (10)	17	399	7	13	–	–	–	
	Chloroform	CLO-INT (0)	15.2	224	484	–	−62.6	52.5	–	[190]
		CLO-INT)(1)	14.9	225	500	–	−62.5	53.9	–	
		CLO-INT (3)	16.4	295	517	–	−62.7	51.5	–	
		CLO-INT (5)	16.6	307	496	–	−62.4	53.4	–	
ESO	MTHPA/1-methylimidazole	OTAC (0)	1.33	2.72	78	–	30.3	–	–	[27]
		OTAC (1)	1.52	3.22	71	–	29.6	–	–	
		OTAC (3)	2.18	4.83	74	–	31.6	–	–	
		OTAC (5)	2.90	9.73	80	–	33.6	–	–	
		OTAC (8)	2.87	8.30	76	–	36.3	–	–	

(continued)

Table 1.11 (continued)

Polymer	Curing agent/catalyst/plasticiser/solvent	Filler type and content (wt%)	Mechanical properties				Thermal poperties			Reference
			Tensile strength (MPa)	Tensile modulus (MPa)	Elongation at break (%)	Impact strength (J/m)	T_g (°C)	T_m (°C)	T_d (°C)	
	TETA	Cloisite 30B (0)	1.20	127	–	–	11.8	–	–	[32]
		Cloisite 30B (5)	2.61	133	–	–	20.7	–	–	
		Cloisite 30B (8)	3.64	151	–	–	19.4	–	–	
		Cloisite 30B (10)	3.54	148	–	–	20.2	–	–	
Starch	Glycerol	Cloisite Na^+ (0)	3.3	29.8	62.6	–	–	–	–	[181]
		Cloisite Na^+ (2)	3.1	29.6	55.1	–	–	–	–	
		Cloisite Na^+ (3)	3.2	150.5	37.6	–	–	–	–	
		Cloisite Na^+ (5)	5.2	195.6	46.8	–	–	–	–	

Table 1.12 Mechanical and thermal properties of biopolymer blend/clay nanocomposites

Polymer	Curing agent/catalyst/ plasticiser/ solvent	Filler type and content (wt%)	Mechanical properties				Thermal poperties		Reference
			Tensile strength (MPa)	Tensile modulus (MPa)	Elongation at break (%)	Impact strength (J/m)	T_g (°C)	T_m (-–°C)	
Diglycidylether bisphenol-S (BPSE)/Hyperbranched polyurethane (HBPU) (100/30)	poly(amido amine)	OMMT (0)	11.23	–	110	100	–	-	[172]
		OMMT (1)	12.94	–	67.45	90	–	–	
		OMMT (3)	14.98	–	55.76	87	–	–	
		OMMT (5)	19.78	–	50.14	80	–	–	
Epoxidised M. ferrea L. seed oil/Bisphenol-A (50/50)	poly(amido amine)	OMMT (0)	7.16	–	16.21	–	–	–	[54]
		OMMT (1)	10.73	–	64.12	–	–	–	
		OMMT (2.5)	11.18	–	59.26	–	–	–	
		OMMT (5)	11.4	–	96.4	–	–	–	
HBPU/Bisphenol-A resin (20/80)	poly(amido amine)	OMMT (0)	30.2	–	510	92	−41	50	[243]
		OMMT (1)	37.02	–	465	98	−33	52	
		OMMT (2.5)	40.63	–	457	99	−31	54	
		OMMT (5)	35.1	–	459	97	−35	55	
PCL/polystyrene (PS)	–	Cloisite 15A (3)	31.4	4090	1.21	–	–	–	[244]
	–	Cloisite 30B (3)	41.2	4970	1.21	–	–	–	
PCL/high impact polystyrene (HIPS)	–	Cloisite 15A (3)	21.2	2690	1.06	–	–	–	
	–	Cloisite 30B (3)	19.7	3030	0.84	–	–	–	

(continued)

Table 1.12 (continued)

Polymer	Curing agent/catalyst/ plasticiser/ solvent	Filler type and content (wt%)	Mechanical properties				Thermal poperties		Reference
			Tensile strength (MPa)	Tensile modulus (MPa)	Elongation at break (%)	Impact strength (J/m)	T_g (°C)	T_m (- – °C)	
PCL/PP	–	Cloisite 15A (3)	23.1	2300	3.39	–	–	–	
	–	Cloisite 30B (3)	23	2690	2.45	–	–	–	
PLA/PCL	Chloroform/methanol	HNT (0)	–	–	–	–	45.62	154.47	[211]
		HNT (2)	–	–	–	–	46.22	154.41	
	Dichloromethane/N,N, dimethylformamide	HNT (0)	–	–	–	–	46.43	153.93	
		HNT (2)	–	–	–	–	44.13	153.89	
PLA/PBSA	-	Cloisite 25A (0)	–	1075.2	71.8	–	–	188.2	[180]
		Cloisite 25A (2)	–	1264.6	4.4	–	–	–	
		Cloisite 25A (5)	–	1616.6	4.1	–	–	–	
		Cloisite 25A (10)	–	1940.1	3.6	–	–	–	
		Twice functionalised organoclay (TFC) (2)	–	1407.9	75.5	–	–	186.4	

(continued)

Table 1.12 (continued)

Polymer	Curing agent/catalyst/ plasticiser/ solvent	Filler type and content (wt%)	Mechanical properties				Thermal poperties		Reference
			Tensile strength (MPa)	Tensile modulus (MPa)	Elongation at break (%)	Impact strength (J/m)	T_g (°C)	T_m (-– °C)	
		TFC (5)	–	1624.6	100.6	–	–	185.2	
		TFC (10)	–	1990.3	118.1	–	–	182.3	
PLA/ENR (0)	-	HNT (2)	53.4	1800	3.8	9.9	–	–	[202]
PLA/ENR (5)			41.9	1600	4.3	12	–	–	
PLA/ENR (10)			36.5	1400	2.4	19.6	–	–	
PLA/ENR (15)			24.1	1400	2.2	43.6	–	–	
PLA/ENR (20)			16.9	1200	1.7	36.6	–	–	
PLA/starch		MMT (0)	43	1037	1.3	–	–	–	[210]
		MMT (1.96)	61	1210	1.5	–	–	–	
	-	MMT (3.85)	61	1388	2.4	–	–	–	
		MMT (5.66)	67	998	2.7	–	–	–	
		MMT (7.41)	68	845	2.6	–	–	–	

that the flexural modulus of bionanocomposites increased by 30% with the addition of 4 vol% MMT nanoclays. However, there were no significant changes in flexural strength, T_g and thermal stability. It was also confirmed that nanocomposite morphology was a mix of intercalated and exfoliated structures.

Das and Karak [54] mixed a Mesua Ferrea L. seed oil as the epoxidiser with bisphenol-A based epoxy to manufacture nanocomposites. The epoxidiser worked more like a plasticiser to reduce the viscosity of epoxy resin and enhance the material performance of poly (amidoamine) as the curing agent. However, the curing time was prolonged by increasing the amounts of seed oils, signifying the importance of its mix with epoxy resin rather than acting as a sole biopolymer. The tensile strength increased by 4% at the nanoclay content of 2.5 wt% as opposed to that of neat epoxy. The presence of 50 wt% seed oil yielded a significant increase in elongation at break by approximately 1200% higher than that of neat epoxy, which is ascribed to the better diffusivity of epoxy matrices into nanoclay interlayer areas with the aid of seed oils.

On the other hand, Haq et al. [242] analysed different synergetic behaviours for epoxidised soybean oil and OMMT nanoclays. The addition of 10 wt% epoxidiser with various nanoclay loadings improved nanocomposite toughness by 75% when compared with that of neat unsaturated polyester. Besides, the manufacturing technique and compatibility of epoxidiser and organoclays were also suggested to improve the stiffness and toughness of bionanocomposites with large amounts of biopolymer matrices and organoclays.

1.5.2 Thermal Properties

Thermal properties of biopolymer/clay nanocomposites are summarised in Tables 1.11 and 1.12. Uyama et al. [168] investigated the thermal properties of plant oil based biopolymer/MMT nanoclays nanocomposites, which were cured by latent acid catalyst benzylsulfonium hexafluoroantimonate. The T_g of nanocomposites was reported to increase with increasing the nanoclay content via dynamic mechanical analysis (DMA). For instance, the addition of 5 wt% Cloisite 30B MMT nanoclays increased the T_g of nanocomposite up to 8.9 °C as opposed to that of neat biopolymer.

Lu et al. [167] used different types of functionalised triglycerides soybean oil based biopolymers (i.e. acrylated epoxidised soybean oil (AESO), maleinised acrylated epoxidised soybean oil (MAESO) and soybean oil pentaerythritol maleates (SOPERMA)) with MMT clays to prepare ESO based nanocomposites. The T_g of nanocomposites was found to be influenced by triglyceride soybean oil type. The T_g of MAESO appeared to be higher at 130 °C in contrast with those of the other types of epoxidised soybean oil, namely AESO at 69 °C and SOPERMA at 126 °C with the clay inclusion of 3 wt%.

In addition, Liu et al. [32] selected triethyl tetramine (TETA) as a typical curing agent for the preparation of ESO/Cloisite 30B MMT clay nanocomposites. Increasing the nanoclay content up to 8 wt% appeared to improve the T_g of nanocomposites by about 64% relative to that of neat bioepoxy. Nevertheless, when the content of curing agent reached 1.37 times comparable to that of epoxy, the T_g increased 10 °C as opposed to that of epoxy/curing agent with a mix ratio of 1:1.

Albayrak et al. [31] modified AESO based styrene nanocomposites, resulting in higher thermal properties and better dynamic mechanical properties when compared to those of styrene. The addition of only 2 wt% modified nanoclays was found to increase the storage modulus of nanocomposites by 400%.

1.6 Biodegradability

The biodegradation of biopolymer nanocomposites is defined as the deterioration of physical and chemical properties when exposed to natural, simulated or artificial environments such as soil, compost, sludge, water or sewage [223]. In addition, Chin and Uematsu [245] suggested the biodegradation process of biopolymers happened by the absorption of microorganisms and enzymes in order to break up polymeric molecular chains and reduce their average MWs.

In addition, a number of testing methods listed in Table 1.13 are current laboratory-scaled methods that are widely used for the biodegradability analysis of plastic materials. Biodegradability analysis can be prepared by analytical techniques, oxygen consumption, microbiological, enzymatic and molecular techniques [223]. Morphological or physical appearance of samples, surface morphology, gravimetric analysis, physical and thermal properties, Fluorescence and UV–vis, gas chromatography and respirometry are categorised as typical analytical techniques for determining the degradation of polymers. Another method is oxygen consumption that can quantify their degradation by recording the generated CO_2 in microbial respiration. Besides, microbiological technique is prepared by placing polymer samples into compost, soil, water, specific bacteria or fungi. Meanwhile, enzymatic technique is conducted by adding some enzymatic catalysts to break down molecular bonds in polymeric molecules. The last one is a molecular technique that is implemented in their designated natural environments.

A number of studies of biodegradability of biopolymers and their nanocomposites were conducted with respect to the effect of curing agent. Chow et al. [246] investigated ESO biodegradability cured by MHHPA using a composting method. The effect of different 2-ethyl-4-methyl-imidazole latent catalyst contents was investigated. The additional 0.5 phr of 2-ethyl-4-methyl-imidazole resulted in the highest degradation level of almost 100% after 8-month soil burial tests. In addition, the reduction in ESO biodegradability reached about 40% for 1 phr catalyst, and further decreased below 5% for 2 phr catalyst.

Tsujimoto et al. [247] investigated ESO biodegradability cured by 3- glycidoxypropyltrimethoxysilane (GPTMS) acid catalyst with a biochemical oxygen demand (BOD) method in the activated sludge environment. It was found that the degradability level of ESO/GPTMS could be reached up to 50% after 60 days. By using a similar method, Takahashi et al. [248] studied the effect of different

Table 1.13 Summary of laboratory testing methods for biodegradation

Characterisation techniques	Property analysis
Gravimetric analysis	% Weight loss
Tensile tests	Mechanical strength, modulus, and elongation
XRD/WAXRD	Crystallinity, interlayer spacing and dispersion degree of nanoscale fillers
Differential scanning calorimetry (DSC)	Crystallisation and melting behaviour
Thermogravimetric analysis (TGA)	Thermal stability
DMA	Thermal transitions
Oxygen gas permeability	Oxygen gas transmission rate
UV–visible spectrometry	Transmittance of UV and visible light
Fluorescence microscopy	Enumeration and viability of bacterial cells
Atomic force microscopy (AFM)	Particle size and distribution, surface morphology and microstructure, degree of crystallisation
Fourier transform infrared (FTIR) analysis	Chemical analysis, crystallisation and orientation of polymer, interfacial interactions
NMR	Chemical analysis
Gel permeation chromatography (GPC) Size-exclusion chromatography (SEC)	Average molecular weight
SEM	Surface morphology
TEM	Internal structure and morphology of nanofillers in composites
Clear-zone	Resistance or susceptibility of plastics to microbial degradation
Pour plate or Streak plate	Enumeration and isolation of degrading microorganisms
Turbidity	Estimation of microbial growth and abundance
Enzymatic analysis	Weight loss and enzymatic activity in the liquid medium containing test material and enzymes
Molecular analysis	Detection, identification of degrading microorganisms

curing agents on ESO biodegradability. Four different curing agents were used including terpene-based acid anhydride (TPAn), maleinated linseed oil (LOAn), a thermally latent cationic polymerisation catalyst (CPI) and hexahydrophthalic anhydride (HPAn). The biodegradability of ESO/LOAn was found to be the highest as compared with other curing agents (CPI, TPAn and HPAn), which was around 20% in 1 month compared to approximately 10% for the other types of curing agents.

On the other hand, biodegradation standard testing methods of biopolymers for home and industry-based applications are listed in Table 1.14. ASTM D6400 and EN13432 are the common method used in material and manufacturing industries while aerobic composting is based on ISO 14855–1 and ISO 14855–2. Such testing standards vary from country to country with ASTM D5338-15 for the USA, CEN

Table 1.14 List of biodegradation testing standards

Standard	Description
AS 4736–2006	Biodegradable plastics—Biodegradable plastics suitable for composting and other microbial treatment
ASTM D5338-15	Standard test method for determining aerobic biodegradation of plastic materials under controlled composting conditions, incorporating thermophilic temperatures
ASTM D5929-96 (2009)	Standard test method for determining the biodegradability of materials exposed to municipal solid waste composting conditions by compost respirometry
ASTM D6400-12	Standard specification for labelling of plastics designed to be aerobically composted in municipal or industrial facilities
ASTM D6691-09	Standard test method for determining aerobic biodegradation of plastic materials in the marine environment by a defined microbial consortium or natural sea water inoculum
ASTM D7475-11	Standard test method for determining the aerobic degradation and anaerobic biodegradation of plastic materials under accelerated bioreactor landfill conditions
EN 13,432:2000	Requirements for packaging recoverable through composting and biodegradation—Test scheme and evaluation criteria for the final acceptance of packaging
EN 14,045:2003	Packaging—Evaluation of the disintegration of packaging materials in practical oriented tests under defined composting conditions
EN 14,046:2003	Evaluation of ultimate aerobic biodegradability and disintegration of packaging materials under controlled composting conditions—Method by analysis of released carbon dioxide
EN 14,047:2003	Packaging—Ultimate aerobic biodegradability of packaging materials in an aqueous medium - Method by analysis of evolved carbon dioxide
EN 14,806:2005	Packaging—Preliminary evaluation of the disintegration of packaging materials under simulated composting conditions in a laboratory scale test
ISO 14855–1	Determination of ultimate aerobic biodegradability of plastic materials under controlled composting conditions—Method by analysis of evolved carbon dioxide-Part 1: General method
ISO 14855–2	Determination of ultimate aerobic biodegradability of plastic materials under controlled composting conditions—Method by analysis of evolved carbon dioxide-Part 2: Gravimetric measurement of carbon dioxide evolved in a laboratory-scale test
ISO 20200	Plastics—Determination of the degree of disintegration of plastic materials under simulated composting condition in a laboratory-scale test

EN ISO 14855 for European countries, DIN EN ISO 14855 for Netherlands, JIS K 6953 for Japan, as well as AS4737 for Australia.

1.6.1 Barrier and Water Absorption Properties

Bionanocomposites in industrial and structural applications are quite limited. The hydrophilic characteristic of biopolymer and wettability of clay nanofillers in matrices have proven their capability to improve the barrier properties of biopolymers against oxygen, hydrogen, nitrogen, helium, carbon dioxide, water vapour and reduce the water uptake percentage [249]. Barrier and water uptake properties of polymer/clay nanocomposites are controlled by several factors, consisting of filler features such as volume fraction and aspect ratio of fillers, nanoclay dispersion status like agglomeration, orientation of fillers and free volumes and intrinsic barrier property of polymers. Nielsen theory [250] can be applied to predict the permeability of polymer/clay nanocomposites. This theory focuses on a tortuous path surrounding clay platelet layers (Fig. 1.11). The fundamental principle is to increase the path length influenced by good clay dispersion with high filler aspect ratios and filler contents in nanocomposite systems. In addition, Fredrickson and Bicerano [251] proposed an analytical model for barrier properties of oriented cubic composites to advance the previous study in terms of the effect of disorder orientation and degree of wettability of fillers. Nonetheless, Gusev and Lusti [252] and Gusev [253] also introduced a theoretical study on barrier properties of nanocomposites with different filler shapes. The numerical simulation method and three-dimensional computer model of randomly dispersed fillers were examined. Moreover, current development and theoretical analysis of barrier properties of nanocomposites have been addressed by Xu et al. [254], Choudalakis and Gotsis [255], Ray [256], Kalendova et al. [257], Cui et al. [258], Tan and Thomas [259], etc.

Additionally, studies in relation to barrier properties of polymer/clay nanocomposites with different types of polymers and various clay loadings are listed in Table 1.16. Picard et al. [261] confirmed a decreasing trend of permeability for annealed

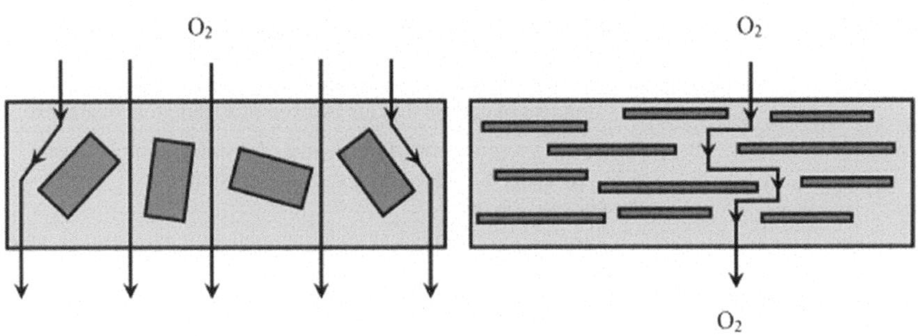

Fig. 1.11 Illustration of tortuous diffusion path in polymer/clay nanocomposites [260]

PLA nanocomposite films reinforced with 4 wt% MMT clays as opposed to amorphous PLA films alone, which is most likely due to the respective contribution from clays and polymer crystalline phase. Sabet et al. [262] reported that barrier properties of PLA and PLA/PCL nanocomposites improved more remarkably towards oxygen molecules when incorporated with DK2 nanoclays and Polybond 3200 as a high-molecular-weight compatiliser as opposed to those based upon low-molecular-weight compatibiliser Licocene. This finding is ascribed to the presence of longer and better dispersed silicate layers in the morphological structures of nanocomposites compatibilised by Polybond 3200. On the other hand, Gain et al. [263] investigated PLC/MMT nanocomposites and found that their highest barrier properties were derived from exfoliated clay dispersion and the highest grating density. It is the similar case for the permeability and diffusion coefficients. Oguzlu and Tihminlioglu [264] studied the barrier properties of chitosan/clay nanocomposite films, which indicated that water vapour and oxygen permeabilities were reduced with the addition of clay particles owing to the increased tortuous path with increasing the clay content. In addition, Nielsen model offered the best fit to the experimental permeability data obtained in this study. Swain et al. [99] prepared soy protein-based nanocomposites by solution intercalation with different loadings of organoclay fillers. The results showed that the significant improvement of oxygen barrier properties was achieved with increasing the clay content along with higher T_g values.

Other studies focused on water uptake (i.e. moisture absorption) feature of polymer/clay nanocomposites shown in Table 1.15 as well. Rhim et al. [265] found

Table 1.15 Barrier properties of polymer/clay nanocomposites

Polymer	Type of clay (wt%)	Permeant	Permeability ($m^2/s{\cdot}Pa$)	Reduction (%)	References
PLA	OMMT (7.9)	O_2	2.85×10^{-18}	24	[261]
PLA	OMMT (7.9)	He	6.79×10^{-17}	23	[261]
PLA/PCL	OMMT (3)	O_2	6.39×10^{-19}	26	[262]
PCL	OMMT (3)	CO_2	3.28×10^{-17}	68	[263]
Polymer	Type of clay (wt%)	Permeant	Permeability ($g/m.s{\cdot}Pa$)	Reduction (%)	References
PLA	Cloisite Na^+(5)	Water	-	-	[265]
	Cloisite 20A(5)	Water	-	5	
	Cloisite 30B(5)	Water	-	36	
PCL	Cloisite Na^+(5)	Water	1.79×10^{-11}	-	[188]
	Cloisite 20A(5)	Water	1.30×10^{-11}	-	
	Cloisite 30B(5)	Water	1.21×10^{-11}	-	
Starch	MMT(3)	Water	3.94×10^{-12}	-	[266]
	OMMT(3)	Water	4.52×10^{-12}	-	
Epoxy	KH-MT(5)	Water	-	57	[267]
	Cloisite 20A(5)	Water	-	66	
	Nanoclay I30E(5)	Water	-	86	

Note KH-MT denotes quaternary alkylamine-modified MMT clays

that tensile, water vapour barrier and antimicrobial properties of PLA/clay nanocomposite films varied with respect to nanoclay type and concentration. In particular, the addition of nanoclays benefited the enhanced barrier properties against water vapour transmission with a great potential for corresponding nanocomposite films to be used in food and beverage packaging. Ludueña et al. [188] demonstrated that PCL/Cloisite 20A clay nanocomposites possessed best clay dispersion, resulting in water vapour permeability being decreased by 40% as a good candidate to replace commodity polymers in disposable polymer packaging applications. Moreover, Tang et al. [266] reported that better water vapour barrier properties were obtained from starch/MMT clay composites as opposed to starch/I30E organomodified clay composites and pristine starch, which was evidenced by lower water vapour permeability (WVP) in a range of 22–67% when increasing the clay content. Kim et al. [267] showed that the moisture permeability of epoxy/clay nanocomposites was decreased with increasing the clay content in good accordance with predicted results based on a simple tortuous path model. This phenomenon arose from large clay aspect ratio to increase an effective penetration path for the permeability reduction. Alamri and Low [268] investigated the effect of moisture absorption on mechanical properties of epoxy nanocomposites reinforced with OMMTs, HNTs, nano-silicon carbide (n-SiC). Their study demonstrated that the addition of those nanofillers at 5 wt% to epoxy resins reduced the moisture absorption by 25, 20 and 33%, respectively, as well as the diffusivity by 30, 32 and 36% accordingly for corresponding composite materials when compared with those of unfilled epoxy. In addition, the flexural strengths decreased by about 13, 18 and 12% for nanocomposites reinforced with 5 wt% of OMMTs, HNTs and n-SiC, respectively after 130 days for water exposure. However, the impact strengths of nanocomposites appeared to conversely increase by about 21, 16 and 19% after the water immersion for nanocomposites reinforced with 1 wt% OMMTs, HNTs and n-SiC, respectively. Vlasveld et al. [269] studied the moisture absorption of PA6/Cloisite 30B clay nanocomposites. It was revealed that water absorption of PA6 nanocomposites decreased by about 30% with increasing the clay content up to 10.6 wt% as opposed to that of neat PA6.

1.7 Summary

Biopolymer/clay nanocomposites can be prepared by mixing conventional polymers and/or biopolymers with clay nanofillers, compatibilisers and solvents. The materials were well tailored to fabricate better biopolymer/clay nanocomposites by improving the compatibility between biopolymer matrices and clay nanofillers, optimising processing parameters and methods including clay type and modification, clay surface chemistry, biopolymer types, catalysts, and/or curing agent types, interaction of compatibilisers with polymer matrices. The successful determination of these parameters plays an important role in final well-tailored material performance including morphological structures, mechanical, thermal and biodegradable properties of biopolymer/clay nanocomposites.

References

1. Mittal V (2011) Bio-nanocomposites: Future High-value Material. In: Mittal V (Ed), Nanocomposites with biodegradable polymers: Synthesis, properties, and future perspectives. Oxford University Press: Oxford, 1–27.
2. Mülhaupt R (2013) Green polymer chemistry and bio-based plastics: Dreams and reality. Macromol Chem Phys 214(2): 159–174.
3. Ray SS, Bousmina M (2005) Biodegradable polymers and their layered silicate nanocomposites: In greening the 21st century materials world. Prog Mater Sci 50(8): 962–1079.
4. Reddy MM, Vivekanandhan S, Misra M, Bhatia SK, Mohanty AK (2013) Biobased plastics and bionanocomposites: Current status and future opportunities. Prog Polym Sci 38(10–11): 1653–1689.
5. Babu RP, O'Connor K, Seeram R (2013) Current progress on bio-based polymers and their future trends. Prog Biomater 2(1): 8.
6. Ratna D, Banthia AK (2000) Epoxidised soybean oil toughened epoxy adhesive. J Adhes Sci Technol 14(1): 15–25.
7. Tan SG, Chow WS (2011) Curing characteristics and thermal properties of epoxidised soybean oil based thermosetting resin. J Am Oil Chem Soc 88(7): 915–923.
8. Park S-J, Jin F-L, Lee J-R (2004) Effect of biodegradable epoxidised castor oil on physicochemical and mechanical properties of epoxy resins. Macromol Chem Phys 205(15): 2048–2054.
9. Jaillet F, Desroches M, Auvergne R, Boutevin B, Caillol S (2013) New biobased carboxylic acid hardeners for epoxy resins. Eur J Lipid Sci Technol 115(6): 698–708.
10. Alexandre M, Dubois P (2000) Polymer-layered silicate nanocomposites: Preparation, properties and uses of a new class of materials. Mater Sci Eng R: Rep 28(1–2):1–63.
11. Kornmann X, Lindberg H, Berglund LA (2001) Synthesis of epoxy–clay nanocomposites: influence of the nature of the clay on structure. Polymer 42(4):1303–1310.
12. Dong Y, Chaudhary D, Ploumis C, Lau K-T(2011) Correlation of mechanical performance and morphological structures of epoxy micro/nanoparticulate composites. Compos Part A: Appl Sci Manuf 42(10):1483–1492.
13. Auvergne R, Caillol S, David G, Boutevin B, Pascault J-P (2013) Biobased thermosetting epoxy: Present and future. Chem Rev 114(2):1082–1115.
14. Khot SN, Lascala JJ, Can E, Morye SS, Williams GI, Palmese GR, Kusefoglu SH, Wool RP (2001) Development and application of triglyceride-based polymers and composites. J Appl Polym Sci 82(3): 703–723.
15. Gupta AP, Ahmad S, Dev A (2010) Development of novel bio-based soybean oil epoxy resins as a function of hardener stoichiometry. Polym-Plast Technol Eng 49(7): 657–661.
16. Zhang J, Hu S, Zhan G, Tang X, Yu Y (2013) Biobased nanocomposites from clay modified blend of epoxidised soybean oil and cyanate ester resin. Prog Org Coat 76(11): 1683–1690.
17. Tan SG, Ahmad Z, Chow WS (2014) Interpenetrating polymer network structured thermosets prepared from epoxidised soybean oil/diglycidyl ether of bisphenol A. Polym Int 63(2):273–279.
18. Samper MD, Petrucci R, Sanchez-Nacher L, Balart R, Kenny JM (2015) Properties of composite laminates based on basalt fibers with epoxidised vegetable oils. Mater Des 72: 9–15.
19. Mauck SC, Wang S, Ding W, Rohde BJ, Fortune CK, Yang G, Ahn, S-K, Robertson ML (2016) Biorenewable tough blends of polylactide and acrylated epoxidised soybean oil compatibilised by a polylactide star polymer. Macromolecules 49(5):1605–1615.
20. Niedermann P, Szebényi G, Toldy A (2014) Effect of epoxidised soybean oil on curing, rheological, mechanical and thermal properties of aromatic and aliphatic epoxy resins. J Polym Environ 22(4): 525–536.
21. Shabeer A, Sundararaman S, Chandrashekhara K, Dharani LR (2007) Physicochemical properties and fracture behavior of soy-based resin. J Appl Polym Sci 105(2): 656–663.

22. Wang R, Schuman T, Vuppalapati RR, Chandrashekhara K (2014) Fabrication of bio-based epoxy-clay nanocomposites. Green Chem 16(4): 1871–1882.
23. Haq M, Burgueño R, Mohanty AK, Misra M (2008) Hybrid bio-based composites from blends of unsaturated polyester and soybean oil reinforced with nanoclay and natural fibers. Compos Sci Technol 68(15–16): 3344–3351.
24. Shibata M, Nakai K (2010) Preparation and properties of biocomposites composed of bio-based epoxy resin, tannic acid, and microfibrillated cellulose. Polym Sci Part B: Polym Phys 48(4): 425–433.
25. Torres JP, Hoto R, Andrés J, García-Manrique JA (2013) Manufacture of green-composite sandwich structures with basalt fiber and bioepoxy resin. Adv Mater Sci Eng (2013) 214506.
26. Usuki A, Kojima Y, Kawasumi M, Okada A, Fukushima Y, Kurauchi T, Kamigaito O (1993) Synthesis of nylon 6-clay hybrid. J Mater Res 8(5): 1179–1184.
27. Tanrattanakul V, Saithai P (2009) Mechanical properties of bioplastics and bioplastic–organoclay nanocomposites prepared from epoxidised soybean oil with different epoxide contents. J Appl Polym Sci 114(5): 3057–3067.
28. Takada Y, Shinbo K, Someya Y, Shibata M (2009) Preparation and properties of bio-based epoxy montomorillonite nanocomposites derived from polyglycerol polyglycidyl ether and ε-polylysine. J Appl Polym Sci 113(1): 479–484.
29. Sermsantiwanit K, Phattanarudee S (2012) Preparation of bio-based nanocomposite emulsions: Effect of clay type. Prog Org Coat 74(4): 660–666.
30. Shibata M (2013) Bio-based epoxy resin/clay nanocomposites. In: Mittal V (ed.) Thermoset Nanocomposites. Wiley: Weinheim, 189–209.
31. Albayrak Ö, Şen S, Çaylı G, Ortaç B (2013) Bio-based polymer nanocomposites based on layered silicates having a reactive and renewable intercalant. J Appl Polym Sci 130(3): 2031–2041.
32. Liu ZS, Erhan SZ, Xu J (2005) Preparation, characterisation and mechanical properties of epoxidised soybean oil/clay nanocomposites. Polymer 46(23):10119–10127.
33. Bourbigot S, Fontaine G, Bellayer S, Delobel R (2008) Processing and nanodispersion: A quantitative approach for polylactide nanocomposite. Polym Test 27(1): 2–10.
34. Haq M, Burgueño R, Mohanty AK, Misra M (2009) Processing techniques for bio-based unsaturated-polyester/clay nanocomposites: Tensile properties, efficiency, and limits. Compos Part A: Appl Sci Manuf 40(4): 394–403.
35. Chivrac F, Pollet E, Dole P, Avérous L (2010) Starch-based nano-biocomposites: Plasticiser impact on the montmorillonite exfoliation process. Carbonhydr Polym 79(4), 941–947.
36. Ojijo V, Ray SS (2013) Processing strategies in bionanocomposites. Prog Polym Sci 38(10–11):1543–1589.
37. Wu L, Hoa SV, Minh T, Ton T (2006) Effects of composition of hardener on the curing and aging for an epoxy resin system. J Appl Polym Sci 99(2): 580–588.
38. Liu L, Li M (2010) Curing mechanisms and kinetic analysis of DGEBA cured with a novel imidazole derivative curing agent using DSC techniques. J Appl Polym Sci 117(6): 3220–3227.
39. Wypych F, Satyanarayana KG (2005) Functionalisation of single layers and nanofibers: a new strategy to produce polymer nanocomposites with optimised properties. J Colloid Interface Sci 285(2): 532–543.
40. Dong Y, Bhattacharyya D (2008) Effects of clay type, clay/compatibiliser content and matrix viscosity on the mechanical properties of polypropylene/organoclay nanocomposites. Compos Part A: Appl Sci Manuf 39(7): 1177–1191.
41. Alzina C, Mija A, Vincent L, Sbirrazzuoli N (2012) Effects of incorporation of organically modified montmorillonite on the reaction mechanism of epoxy/amine cure. J Phys Chem B 116: 5786–5794.
42. Nakamura R, Netravali AN, Hosur MV (2012) Effect of halloysite nanotube incorporation in epoxy resin and carbon fiber ethylene/ammonia plasma treatment on their interfacial property. J Adhes Sci Technol 26(8–9):1295–1312.

43. Dong Y, Bhattacharyya D (2012) Investigation on the competing effects of clay dispersion and matrix plasticisation for polypropylene/clay nanocomposites. Part I: morphology and mechanical properties. J Mater Sci 47(8): 3900–3912.
44. Bur AJ, Lee YH, Roth SC, Start PR (2005) Measuring the extent of exfoliation in polymer/clay nanocomposites using real-time process monitoring methods. Polymer 46(24): 10908–10918.
45. Wang H, Hoa SV, Wood-Adams PM (2006) New method for the synthesis of clay/epoxy nanocomposites. J Appl Polym Sci 100(6): 4286–4296.
46. Barbas JM, Machado AV, Covas JA (2013). Evolution of dispersion along the extruder during the manufacture of polymer–organoclay nanocomposites. Chem Eng Sci 98: 77–87.
47. Kaboorani A, Riedl B, Blanchet P (2013) Ultrasonication technique: A method for dispersing nanoclay in wood adhesives. J Nanomater 2013: 341897.
48. Zeng Q, Yu A (2010) Prediction of the mechanical properties of nanocomposites. In Mittal V (ed) Optimisation of Polymer Nanocomposite Properties. Wiley: Weinheim, 301–331.
49. Dong Y, Bhattacharyya D (2010) Mapping the real micro/nanostructures for the prediction of elastic moduli of polypropylene/clay nanocomposites. Polymer 51(3): 816–824.
50. Brune DA, Bicerano J (2002) Micromechanics of nanocomposites: comparison of tensile and compressive elastic moduli, and prediction of effects of incomplete exfoliation and imperfect alignment on modulus. Polymer 43(2): 369–387.
51. Miyagawa H, Foo KH, Daniel IM, Drzal LT (2005) Mechanical properties and failure surface morphology of amine-cured epoxy/clay nanocomposites. J Appl Polym Sci 96(2): 281–287.
52. Alateyah AI, Dhakal HN, Zhang ZY (2013) Processing, properties and applications of polymer nanocomposites based on layer silicates: A review. Adv Polym Technol 32(4) 21368.
53. Bensadoun F, Kchit N, Billotte C, Bickerton S, Trochu F, Ruiz E (2011) A study of nanoclay reinforcement of biocomposites made by liquid composite molding. Int J Polym Sci 2011: 964193.
54. Das G, Karak N (2009) Epoxidised Mesua ferrea L. seed oil-based reactive diluent for BPA epoxy resin and their green nanocomposites. Prog Org Coat 66(1): 59–64.
55. Atiemo-Obeng VA, Penney W, Armenante PM (2004) Solid-liquid mixing. In: Paul EL, Atiemo-Obeng VA, Kresta SM (Eds.) Handbook of industrial mixing: Science and practice. Wiley: Hoboken, 543–584.
56. Zini E, Scandola M (2011) Green composites: An overview. Polym Compos 32(12): 1905–1915.
57. Avérous L (2004) Biodegradable multiphase systems based on plasticised starch: A review. J Macromol Sci Part C: Polym Rev 44(3): 231–274.
58. Rüsch gen Klaas M, Warwel S (1999) Complete and partial epoxidation of plant oils by lipase-catalysed perhydrolysis. Ind Crops Prod 9(2):125–132.
59. Tan SG, Chow WS (2010) Biobased epoxidised vegetable oils and its greener epoxy blends: A review. Polym-Plast Technol Eng 49(15):1581–1590.
60. Hernandez N, Williams RC, Cochran EW (2014) The battle for the "green" polymer. Different approaches for biopolymer synthesis: bioadvantaged vs. bioreplacement. Org Biomol Chem 12(18): 2834–2849.
61. Espinoza-Perez JD, Nerenz BA, Haagenson DM, Chen Z, Ulven CA, Wiesenborn DP (2011) Comparison of curing agents for epoxidised vegetable oils applied to composites. Polym Compos 32(11) 1806–1816.
62. Park S-J, Jin F-L, Lee J-R, Shin J-S (2005) Cationic polymerization and physicochemical properties of a biobased epoxy resin initiated by thermally latent catalysts. Euro Polym J 41(2): 231–237.
63. España JM, Sánchez-Nacher L, Boronat T, Fombuena V, Balart R (2012) Properties of biobased epoxy resins from epoxidized soybean oil (ESBO) cured with maleic anhydride (MA). J Am Oil Chem Soc 89: 2067–2075.
64. Chen Y, Xi Z, Zhao L (2016) New bio-based polymeric thermosets synthesized by ring-opening polymerisation of epoxidised soybean oil with a green curing agent. Eur Polym J 84: 435–447.

65. Xie F, Pollet E, Halley PJ, Avérous L (2013) Starch-based nano-biocomposites. Prog Polym Sci 38(10–11): 1590–1628.
66. Liu H, Xie F, Yu, L, Chen L, Li L (2009) Thermal processing of starch-based polymers. Prog Polym Sci 34(12) 1348–1368.
67. Rodríguez-Llamazares S, Rivas BL, Pérez M, Perrin-Sarazin F (2012) Poly(ethylene glycol) as a compatibiliser and plasticiser of poly(lactic acid)/clay nanocomposites. High Perform Polym 24(4): 254–261.
68. Jadawi A, Saiter J-M, Barbe F, Negahban M (2014) Porous biodegradable Starch-based polymer: Effects of plasticisers on the physical properties. Macromol Symp 340(1): 65–72.
69. Imre B, Pukánszky B (2013) Compatibilisation in bio-based and biodegradable polymer blends. Eur Polym J 49(6): 1215–1233.
70. Martin O, Avérous L (2001) Poly(lactic acid): Plasticisation and properties of biodegradable multiphase systems. Polymer 42(14): 6209–6219.
71. Avella M, Buzarovska A, Errico ME, Gentile G, Grozdanov A (2009) Eco-challenges of bio-based polymer composites. Materials 2(3): 911–925.
72. Woodruff MA, Hutmacher DW (2010) The return of a forgotten polymer—Polycaprolactone in the 21st century. Prog Polym Sci 35(10): 1217–1256.
73. Winnacker M, Rieger B (2016) Poly(ester amide): Recent insights into synthesis, stability and biomedical applications. Polym Chem 7(46): 7039–7046.
74. Kropp M, Morawa K-M, Mihov G, Salz A, Harmening N, Franken A, Kemp, A, Dias AA, Thies J, Johnen S, Thumann G (2014). Biocompatibility of Poly(ester amide) (PEA) microfibrils in ocular tissues. Polymers 6(1): 243–260.
75. Zou Y, Wang L, Zhang H, Qian Z, Mou L, Wang J, Liu X (2004). Stabilisation and mechanical properties of biodegradable aliphatic polyesteramide and its filled composites. Polym Degrad Stab 83(1): 87–92.
76. Hao Y, Chen M, Zhao J, Zhang Z, Yang W (2013). Synthesis and Properties of Polyesteramides Having Short Nylon-610 Segments in the Main Chains through Polycondensation and Chain Extension. Ind Eng Chem Res 52(19): 6410–6421.
77. Sun H-Y, Yin T, Zhao J-B, Zhang Z-Y, Yang W-T (2013) Synthesis and characterisation of biodegradable polyesteramides constructed mainly by alternating diesterdiamide units from N,N′-bis(2-hydroxyethyl)-adipamide and diacids. Chinese J Polym Sci 31(3): 452–461.
78. Ghosal K, Latha MS, Thomas S (2014) Poly(ester amides) (peas) – scaffold for tissue engineering applications. Eur Polym J. 60: 58–68.
79. Imre B, Renner K, Pukánszky B (2014) Interactions, structure and properties in poly(lactic acid)/thermoplastic polymer blends. Express Polym Lett 8(1): 2–14.
80. Supanchaiyamat N, Shuttleworth PS, Hunt AJ, Clark JH, Matharu AS (2012) Thermosetting resin based on epoxidised linseed oil and bio-derived crosslinker. Green Chem 14(6):1759–1765.
81. Liu D, Qi Z, Zhang Y, Xu J, Guo B (2015) Poly(butylene succinate) (PBS)/ionic liquid plasticised starch blends: Preparation, characterisation, and properties. Starch 67(9-10): 802–809.
82. Park JW, Im SS, Kim SH, Kim YH (2000) Biodegradable polymer blends of poly(L-lactic acid) and gelatinised starch. Polym Eng Sci 40(12): 2539–2550.
83. Avérous L, Moro L, Dole P, Fringant C (2000) Properties of thermoplastic blends: Starch–polycaprolactone. Polymer 41(11): 4157–4167.
84. Avérous L, Fauconnier N, Moro L, Fringant C (2000) Blends of thermoplastic starch and polyesteramide: Processing and properties. J Appl Polym Sci 76(7): 1117–1128.
85. Wang Y, Chiao SM, Hung TF, Yang SY (2012) Improvement in toughness and heat resistance of poly(lactic acid)/polycarbonate blend through twin-screw blending: Influence of compatibiliser type. J Appl Polym Sci 125(S2): E402–E412.
86. Yokohara T, Yamaguchi M (2008) Structure and properties for biomass-based polyester blends of PLA and PBS. Eur Polym J 44(3): 677–685.
87. Park SB, Hwang SY, Moon CW, Im SS, Yoo ES (2010) Plasticiser effect of novel PBS ionomer in PLA/PBS ionomer blends. Macromol Res 18(5): 463–471.

88. Mohamed A, Gordon SH, Biresaw G (2007) Poly(lactic acid)/polystyrene bioblends characterised by thermogravimetric analysis, differential scanning calorimetry, and photoacoustic infrared spectroscopy. J Appl Polym Sci 106(3):1689–1696.
89. Zhang B, Sun B, Bian X, Li G, Chen X (2017) High melt strength and high toughness PLLA/PBS blends by copolymerisation and in situ reactive compatibilisation. Ind Eng Chem Res 56(1): 52–62.
90. Miyagawa H, Mohanty A, Drzal LT, Misra M (2004) Effect of clay and alumina-nanowhisker reinforcements on the mechanical properties of nanocomposites from biobased epoxy: A comparative study. Ind Eng Chem Res 43(22): 7001–7009.
91. Miyagawa H, Mohanty AK, Misra M, Drzal LT (2004) Thermo-physical and impact properties of epoxy containing epoxidised linseed oil, 1. Macromol Mater Eng 289(7): 629–635.
92. Jin F-L, Park S-J (2008) Thermomechanical behavior of epoxy resins modified with epoxidised vegetable oils. Polym Int 57: 577–583.
93. Ratna D (2001) Mechanical properties and morphology of epoxidised soyabean-oil-modified epoxy resin. Polym Int 50(2): 179–184.
94. Altuna FI, Espósito LH, Ruseckaite RA, Stefani PM (2011) Thermal and mechanical properties of anhydride-cured epoxy resins with different contents of biobased epoxidised soybean oil. J Appl Polym Sci 120(2): 789–798.
95. Sahoo SK, Mohanty S, Nayak SK (2015) Toughened bio-based epoxy blend network modified with transesterified epoxidised soybean oil: Synthesis and characterisation. RSC Adv 5(18):13674–13691.
96. Czub P (2006) Application of modified natural oils as reactive diluents for epoxy resins. Macromol Symp 242(1): 60–64.
97. Liu, Z. S., Erhan, S. Z., and Calvert, P. D. (2004). Solid freeform fabrication of epoxidised soybean oil/epoxy composites with di-, tri-, and polyethylene amine curing agents. *Journal of Applied Polymer Science, 93*(1), 356–363.
98. Roudsari GM, Mohanty AK, Misra M (2014) Study of the curing kinetics of epoxy resins with biobased hardener and epoxidised soybean oil. ACS Sustain Chem Eng *2*(9): 2111–2116.
99. Swain SK, Priyadarshini PP, Patra SK (2012) Soy protein/clay bionanocomposites as ideal packaging materials. Polym-Plast Technol Eng 51(12): 1282–1287.
100. Bitinis N, Verdejo R, Bras J, Fortunati E, Kenny JM, Torre L, López-Manchado MA (2013) Poly(lactic acid)/natural rubber/cellulose nanocrystal bionanocomposites Part I. Processing and morphology. Carbohydr Polym 96: 611–620.
101. Pandey JK, Nakagaito AN, Takagi H (2013) Fabrication and applications of cellulose nanoparticle-based polymer composites. Polym Eng Sci 53(1): 1–8.
102. Triantafillidis CS, LeBaron PC, Pinnavaia TJ (2002) Thermoset epoxy–clay nanocomposites: The dual role of α,ω-diamines as clay surface modifiers and polymer curing agents. J Solid State Chem 167(2): 354–362.
103. Mago G, Ray SS, Shofner ML, Wang S, Zhang J (2014) Polymer nanocomposite processing, characterisation and applications 2013. J Nanomater 2014: 403492.
104. Theng BKG (2012) Polymer-clay nanocomposites. In: Theng BKG (Ed), Formation and properties of clay-polymer complex Elsevier: Amsterdam, 201–241.
105. Bergaya F, Detellier C, Lambert JF, Lagaly G (2013) Introduction to clay–polymer nanocomposites (CPN). In Bergaya F, Lagaly G (Eds), Handbook of Clay Science. Elsevier: Amsterdam, 655–677.
106. Tjong SC (2006) Structural and mechanical properties of polymer nanocomposites. Mater Sci Eng R: Rep *53*(3–4): 73–197.
107. Deng S, Zhang J, Ye L, Wu J (2008) Toughening epoxies with halloysite nanotubes. Polymer *49*(23): 5119–5127.
108. Du M, Guo B, Lei Y, Liu M, Jia D (2008) Carboxylated butadiene–styrene rubber/halloysite nanotube nanocomposites: Interfacial interaction and performance. Polymer 49: 4871–4876.
109. Handge UA, Hedicke-Höchstötter K, Altstädt, V (2010). Composites of polyamide 6 and silicate nanotubes of the mineral halloysite: Influence of molecular weight on thermal, mechanical and rheological properties. Polymer 51(12): 2690–2699.

110. Ismail H, Pasbakhsh P, Fauzi MNA, Abu Bakar A (2008) Morphological, thermal and tensile properties of halloysite nanotubes filled ethylene propylene diene monomer (EPDM) nanocomposites. Polym Test 27(7): 841–850.
111. Lecouvet B, Gutierrez JG, Sclavons M, Bailly C (2011) Structure–property relationships in polyamide 12/halloysite nanotube nanocomposites. Polym Degrad Stab 96(2): 226–235.
112. Yah WO, Yamamoto K, Jiravanichanun N, Otsuka H, Takahara A (2010) Imogolite reinforced nanocomposites: Multifaceted green materials. Materials 3(3): 1709–1745.
113. Yamamoto K, Otsuka H, Wada S-I, Sohn D, Takahara A (2005) Preparation and properties of [poly(methyl methacrylate)/imogolite] hybrid via surface modification using phosphoric acid ester. Polymer 46(26): 12386–12392.
114. Nuruzzaman M, Rahman MM, Liu Y, Naidu R (2016) Nanoencapsulation, nano-guard for pesticides: A new window for safe application. J Agr Food Chem 64(7): 1447–1483.
115. Bordes P, Pollet E, Avérous L (2009. Nano-biocomposites: Biodegradable polyester/nanoclay systems. Prog Polym Sci 34(2): 125–155.
116. Xu W-B, Bao S-P, He P-S (2002). Intercalation and exfoliation behavior of epoxy resin/curing agent/montmorillonite nanocomposite. J Appl Polym Sci 84(4): 842–849.
117. Liu M, Guo B, Du M, Cai X, Jia D (2007) Properties of halloysite nanotube–epoxy resin hybrids and the interfacial reactions in the systems. Nanotechnology 18(45): 455703.
118. Lvov Y, Abdullayev E (2013) Functional polymer–clay nanotube composites with sustained release of chemical agents. Prog Polym Sci 38(10–11):1690–1719.
119. Lvov YM, Shchukin DG, Möhwald H, Price RR (2008). Halloysite clay nanotubes for controlled release of protective agents. ACS Nano 2(5): 814–820.
120. Zeng S, Reyes C, Liu J, Rodgers PA, Wentworth SH, Sun L (2014) Facile hydroxylation of halloysite nanotubes for epoxy nanocomposite applications. Polymer 55(25): 6519–6528.
121. Carli LN, Crespo JS, Mauler RS (2011) PHBV nanocomposites based on organomodified montmorillonite and halloysite: The effect of clay type on the morphology and thermal and mechanical properties. Compos Part A: Appl Sci Manuf 42:1601–1608.
122. Liu M, Guo B, Du M, Lei Y, Jia D (2008). Natural inorganic nanotubes reinforced epoxy resin nanocomposites. J Polym Res 15(3):205–212.
123. Tang Y, Deng S, Ye L, Yang C, Yuan Q, Zhang J, Zhao C (2011) Effects of unfolded and intercalated halloysites on mechanical properties of halloysite–epoxy nanocomposites. Compos Part A: Appl Sci Manuf 42(4): 345–354.
124. Tang Y, Ye L, Deng S, Yang C, Yuan W (2012) Influences of processing methods and chemical treatments on fracture toughness of halloysite–epoxy composites. Mater Des 42: 471–477.
125. Dong Y, Chaudhary D, Haroosh H, Bickford T (2011) Development and characterisation of novel electrospun polylactic acid/tubular clay nanocomposites. J Mater Sci 46(18): 6148–6153.
126. Haroosh HJ, Dong Y, Ingram GD (2013) Synthesis, morphological structures, and material characterisation of electrospun PLA:PCL/Magnetic nanoparticle composites for drug delivery. J Polym Sci Part B: Polym Phys 51(22): 1607–1617.
127. Hanid NA, Wahit MU, Guo Q, Mahmoodian S, Soheilmoghaddam M (2014) Development of regenerated cellulose/halloysites nanocomposites via ionic liquids. Carbohydr Polym 99:91–97.
128. Raquez J-M, Habibi Y, Murariu M, Dubois P (2013) Polylactide (PLA)-based nanocomposites. Prog Polym Sci 38(10–11):1504–1542.
129. Nakamura R, Netravali AN, Morgan AB, Nyden MR, Gilman JW (2013). Effect of halloysite nanotubes on mechanical properties and flammability of soy protein based green composites. Fire Mater 37(1):75–90.
130. Abdullayev E, Lvov Y (2010) Clay nanotubes for corrosion inhibitor encapsulation: release control with end stoppers. J Mater Chem 20: 6681–6687.
131. Kelly HM, Deasy PB, Ziaka E, Claffey N (2004) Formulation and preliminary in vivo dog studies of a novel drug delivery system for the treatment of periodontitis. Int J Pharm 274(1–2):167–183.

132. Salam H, Dong Y, Davies I (2015) Development of biobased polymer/clay nanocomposites: a critical review. In: Dong Y, Umer R, Lau AKT (Eds), Fillers and reinforcements for advanced nanocomposites. Woodhead Publishing: Cambridge, 101–132.
133. Imamura S, Hayashi Y, Kajiwara K, Hoshino H, Kaito C (1993) Imogolite: a possible new type of shape-selective catalyst. Ind Eng Chem Res 32(4): 600–603.
134. Tsujimoto Y, Yoshida A, Kobayashi M, Adachi Y (2013) Rheological behavior of dilute imogolite suspensions. Colloid Surf A: Physicochem Eng Asp 435: 109–114.
135. Farmer VC, Fraser AR, Tait JM (1977. Synthesis of imogolite: a tubular aluminium silicate polymer. J Chem Soc Chem Commun (13): 462–463.
136. Boudot JP (1992) Relative efficiency of complexed aluminum noncrystalline Al hydroxide, allophane and imogolite in retarding the biodegradation of citric acid. Geoderma 52(1–2):29–39.
137. Koenderink GH, Kluijtmans SGJM, Philipse AP (1999) On the synthesis of colloidal Imogolite fibers. J Colloid Interf Sci 216(2): 429–431.
138. Marzán LL, Philipse AP (1994) Synthesis of platinum nanoparticles in aqueous host dispersions of inorganic (imogolite) rods. Colloid Surf A: Physicochem Eng Asp 90(1): 95–109.
139. Barrett SM, Budd PM, Price C (1991) The synthesis and characterisation of imogolite. Eur Polym J 27: 609–612.
140. Lee H, Ryu J, Kim D, Joo Y, Lee SU, Sohn D (2013) Preparation of an imogolite/poly(acrylic acid) hybrid gel. J Collid Interf Sci 406: 165–171.
141. McCutcheon A, Hu J, Kamali Kannangara GS, Wilson MA, Reddy N (2005) 29Si labelled nanoaluminosilicate Imogolite. J Non-Cryst Solids 351(24–26): 1967–1972.
142. Guimaraes L, Pinto YN, Lourenco MP, Duarte HA (2013). Imogolite-like nanotubes: structure, stability, electronic and mechanical properties of the phosphorous and arsenic derivatives. Phys Chem Chem Phys 15(12): 4303–4309.
143. Ma W, Otsuka H, Takahara A, (2011) Poly(methyl methacrylate) grafted imogolite nanotubes prepared through surface-initiated ARGET ATRP. Chem Commu 47(20) 5813–5815.
144. Imamura S, Kokubu T, Yamashita T, Okamoto Y, Kajiwara K, Kanai H (1996) Shape-selective Copper-loaded Imogolite catalyst. J Catal 160(1):137–139.
145. Ma W, Otsuka H, Takahara A (2011) Preparation and properties of PVC/PMMA-g-imogolite nanohybrid via surface-initiated radical polymerization. Polymer 52 (24) 5543–5550.
146. Kang D-Y, Jones CW, Nair S (2011) Modeling molecular transport in composite membranes with tubular fillers. J Membr Sci 381(1–2): 50–63.
147. Ma W, Yah WO, Otsuka H, Takahara A (2012) Application of imogolite clay nanotubes in organic-inorganic nanohybrid materials. J Mater Chem 22 (24): 11887–11892.
148. Kornmann X, Thomann R, Mülhaupt R, Finter J, Berglund L (2002) Synthesis of aminecured, epoxy-layered silicate nanocomposites: the influence of the silicate surface modification on the properties. J Appl Polym Sci 86 (10):2643–2652.
149. Čapková P, Pospíšil M, Valášková M, Měřínská D, Trchová M, Sedláková Z, Weiss Z, Šimoník J (2006) Structure of montmorillonite cointercalated with stearic acid and octadecylamine: Modeling, diffraction, IR spectroscopy. J Colloid Interf Sci 300: 264–269.
150. Ha SR, Ryu SH, Park SJ, Rhee KY (2007) Effect of clay surface modification and concentration on the tensile performance of clay/epoxy nanocomposites. Mater Sci Eng A 448(1–2): 264–268.
151. Li C, Liu J, Qu X, Guo B, Yang Z (2008) Polymer-modified halloysite composite nanotubes. J Appl Polym Sci 110(6): 3638–3646.
152. Park S, Lee Y, Kim B, Lee J, Jeong Y, Noh J, Takahara A, Sohn D (2007). Two-dimensional alignment of imogolite on a solid surface. Chem Commun (28): 2917–2919.
153. Bart JCJ (2006) Plastics Additives: Advanced Industrial Analysis, IOS Press: Amsterdam, 445–596.
154. Machado AV, Barbas JM, Covas JA (2014) Near IR spectroscopy for the characterisation of dispersion in polymer–clay nanocomposites. In: Mittal V (Ed) Synthesis techniques for polymer nanocomposites, Wiley: Weinheim, 241–266.

155. Zhu L, Wool RP (2006) Nanoclay reinforced bio-based elastomers: Synthesis and characterisation. Polymer 47(24): 8106–8115.
156. Tan WL, Salehabadi A, Mohd Isa MH, Abu Bakar M, Abu Bakar NHH (2016) Synthesis and physicochemical characterisation of organomodified halloysite/epoxidised natural rubber nanocomposites: a potential flame-resistant adhesive. J Mater Sci 51(2): 1121–1132.
157. Donescu D, Ianchis R, Petcu C, Purcar V, Nistor CL, Radovici C, Somoghi R, Popo SF, Perichaud A (2013) Study of the solvents influence on the layered silicates-cation polymer hybrids properties. Dig J Nanomater Biostructures 8(4) 1751–1759.
158. Lebaron PC, Wang Z, Pinnavaia TJ (1999) Polymer-layered silicate nanocomposites: an overview. Appl Clay Sci 15(1–2): 11–29.
159. Burgentzle D, Duchet J, Gerard JF, Jupin A, Fillo B (2004) Solvent-based nanocomposite coatings I. Dispersion of organophilic montmorillonite in organic solvents. J Colloid Interf Sci 278: 26–39.
160. Hutchinson JM, Montserrat S, Roman F, Cortes P, Campos L (2006) Intercalation of epoxy resin in organically modified montmorillonite. J Appl Polym Sci 102(4): 3751–3763.
161. Miyagawa H, Mohanty AK, Burgueno R, Drzal LT, Misra M (2006) Characterisation and thermophysical properties of unsaturated polyester-layered silicate nanocomposites. J Nanosci Nanotechnol 6(2): 464–471.
162. Morgan AB, Harris JD (2004) Exfoliated polystyrene-clay nanocomposites synthesised by solvent blending with sonication. Polymer 45(26): 8695–8703.
163. Lu Y, Larock RC (2007) Bio-based nanocomposites from corn oil and functionalised organoclay prepared by cationic polymerisation. Macromol Mater Eng 292(7): 863–872.
164. Dan CH, Kim YD, Lee M, Min BH, Kim JH (2008) Effect of solvent on the properties of thermoplastic polyurethane/clay nanocomposites prepared by solution mixing. J Appl Polym Sci 108(4): 2128–2138.
165. Brown JM, Curliss D, Vaia RA (2000) Thermoset-layered silicate nanocomposites. Quaternary ammonium montmorillonite with primary diamine cured epoxies. Chem Mater 12 (11):3376–3384.
166. Lee C-H, Kato M, Usuki A (2011) Preparation and properties of bio-based polycarbonate/clay nanocomposites. J Mater Chem 21(19): 6844–6847.
167. Lu J, Hong CK, Wool RP (2004) Bio-based nanocomposites from functionalized plant oils and layered silicate. J Polym Sci Part B: Polym Phys 42:1441–1450.
168. Uyama H, Kuwabara M, Tsujimoto T, Nakano M, Usuki A, Kobayashi S (2003) Green nanocomposites from renewable resources: Plant oil−clay hybrid materials. Chem Mater 15(13): 2492–2494.
169. Paluvai NR, Mohanty S, Nayak SK (2015) Synthesis and characterisation of acrylated epoxidised castor oil nanocomposites. Int J Polym Anal Charact 20(4): 298–306.
170. Altuntaş E, Çayli G, Küsefoğlu S, Nugay N (2008) Renewable polymeric nanocomposite synthesis using renewable functionalised soybean-oil-based intercalant and matrix. Des Monomers Polym 11(4): 371–381.
171. Bhuyan S, Sundararajan S, Lu Y, Larock RC (2010) A study of the physical and terminological properties of biobased polymer–clay nanocomposites at different clay concentrations. Wear 268(5–6): 797–802.
172. Das G, Deka H, Karak N (2013) Bio-based sulfonated epoxy/hyperbranched polyurea-modified MMT nanocomposites. Int J Polym Mater Polym Biomater 62(6):330–335.
173. Das G, Karak N (2009) Vegetable oil-based flame retardant epoxy/clay nanocomposites. Polym Degrad Stab 94(11): 1948–1954.
174. Chen G-X, Kim H-S, Yoon J-S (2007) Synthesis and characterisation of poly(butylene succinate)/epoxy group functionalised organoclay. Polym Int 56(9): 1159–1165.
175. Krook M, Albertsson AC, Gedde UW, Hedenqvist MS (2002). Barrier and mechanical properties of montmorillonite/polyesteramide nanocomposites. Polym Eng Sci 42(6): 1238–1246.
176. Krook M, Morgan G, Hedenqvist MS (2005) Barrier and mechanical properties of injection molded montmorillonite/polyesteramide nanocomposites. Polym Eng Sci 45(1): 135–141.

177. Morales-Gámez L, Franco L, Puiggalí J (2011) Thermal stability studies on clay nanocomposites prepared from a degradable poly(ester amide) constituted by glycolic acid and 6-aminohexanoic acid. Thermochim Acta 512(1–2): 142–149.
178. Morales L, Franco L, Casas MT, Puiggalí (2009) Poly(ester amide)/clay nanocomposites prepared by in situ polymerization of the sodium slat of N-chloroacetyl-6-aminohexanoic acid. J Polym Sci Part A: Polym Chem 47(14): 3616–3629.
179. Chiu F-C (2016) Fabrication and characterisation of biodegradable poly(butylene succinate-co-adipate) nanocomposites with halloysite nanotube and organo-montmorillonite as nanofillers. Polym Test 54:1–11.
180. Chen G-X, Yoon J-S (2005) Morphology and thermal properties of poly(L-lactide)/poly(butylene succinate-co-butylene adipate) compounded with twice functionalised clay. J Polym Sci Part B: Polym Phys 43(5): 478–487.
181. Cyras VP, Manfredi LB, Ton-That M-T, Vázquez A (2008) Physical and mechanical properties of thermoplastic starch/montmorillonite nanocomposite films. Carbonhydr Polym 73(1): 55-63.
182. Marras SI, Kladi KP, Tsivintzelis I, Zuburtikudis I, Panayiotou C (2008) Biodegradable polymer nanocomposites: The role of nanoclays on the thermomechanical characteristics and the electrospun fibrous structure. Act Biomater 4(3): 756–765.
183. Lepoittevin B, Devalckenaere M, Pantoustier N, Alexandre M, Kubies D, Calberg C, Jerome R, Dubois P (2002) Poly(ε-caprolactone)/clay nanocomposites prepared by melt intercalation: mechanical, thermal and rheological properties. Polymer 43(14): 4017–4023.
184. Di Maio ED, Iannace S, Di Y, Del Giacomo E, Nicolais L (2003) Heterogeneous bubble nucleation in PCL/clay nanocomposite foams. Plast Rubber Compos 32(7): 313–317.
185. Ludueña LN, Alvarez VA, Vazquez A (2007) Processing and microstructure of PCL/clay nanocomposites. Mater Sci Eng A 460–461: 121–129.
186. Ludueña LN, Vazquez A, Alvarez VA (2008) Crystallisation of polycaprolactone–clay nanocomposites. J Appl Polym Sci 109(5): 3148–3156.
187. Ludueña L, Vázquez A, Alvarez V (2012) Viscoelastic behavior of polycaprolactone/clay nanocomposites. J Compos Mater 46(6): 677–689.
188. Ludueña LN, Vázquez A, Alvarez VA (2013) Effect of the type of clay organo-modifier on the morphology, thermal/mechanical/impact/barrier properties and biodegradation in soil of polycaprolactone/clay nanocomposites. J Appl Polym Sci 128(5): 2648–2657.
189. Wu T, Xie T, Yang G (2009) Preparation and characterisation of poly(ε-caprolactone)/Na^+-MMT nanocomposites. Appl Clay Sci 45(3): 105–110.
190. Miura S, Teramoto N, Shibata M (2016) Nanocomposites composed of poly(ε-caprolactone) and oligocaprolactone-modified imogolite utilising biomimetic chelating method. J Polym Res 23(2):19.
191. Someya Y, Nakazato T, Teramoto N, Shibata M (2004) Thermal and mechanical properties of poly(butylene succinate) nanocomposites with various organo-modified montmorillonites. J Appl Polym Sci 91(3):1463–1475.
192. Shih YF, Wang TY, Jeng RJ, Wu JY, Teng CC (2007) Biodegradable nanocomposites based on poly(butylene succinate)/organoclay. J Polym Environ 15(2): 151–158.
193. Lim S-K, Lee J-J, Jang S-G, Lee S-I, Lee K-H, Choi HJ, Chin I-J (2011) Synthetic aliphatic biodegradable poly(butylene succinate)/clay nanocomposite foams with high blowing ratio and their physical characteristics. Polym Eng Sci 51(7): 1316–1324.
194. Wu W, Cao X, Luo J, He G, Zhang Y (2014) Morphology, thermal, and mechanical properties of poly(butylene succinate) reinforced with halloysite nanotube. Polym Compos 35(5): 847–855.
195. Kim YH, Kwon SH, Choi HJ, Choi K, Kao N, Bhattacharya SN, Gupta RK (2016) Thermal, mechanical and rheological characterisation of Polylactic Acid/Halloysite nanotube nanocomposites. J Macromol Sci Part B 55(7): 680–692.
196. Das K, Ray D, Banerjee I, Bandyopadhyay NR, Sengupta S, Mohanty AK, Misra M (2010) Crystalline morphology of PLA/clay nanocomposite films and its correlation with other properties. J Appl Polym Sci 118(1): 143–151.

197. Krishnamachari P, Zhang J, Lou J, Yan J, Uitenham L (2009) Biodegradable poly(lactic acid)/clay nanocomposites by melt intercalation: A study of morphological, thermal, and mechanical properties. Int J Polym Anal Charact 14 (4):336–350.
198. Ocak Y, Sofuoglu A, Tihminlioglu F, Böke H (2015) Sustainable bio-nano composite coatings for the protection of marble surfaces. J Cult Herit 16(3): 299–306.
199. De Silva RT, Soheilmoghaddam M, Goh K L, Wahit MU, Bee SAH, Chai S-P, Pasbakhsh P (2016) Influence of the processing methods on the properties of poly(lactic acid)/halloysite nanocomposites. Polym Compos 37(3): 861–869.
200. Tham WL, Chow WS, Poh BT, Mohd Ishak ZA (2016) Poly(lactic acid)/halloysite nanotube nanocomposites with high impact strength and water barrier properties. J Compos Mater 50(28): 3925–3934.
201. Tham WL, Poh BT, Mohd Ishak ZA, Chow WS (2016) Transparent poly(lactic acid)/halloysite nanotube nanocomposites with improved oxygen barrier and antioxidant properties. J Therm Anal Calorim 126(3): 1331–1337.
202. Tham WL, Poh BT, Mohd Ishak ZA, Chow WS (2016) Epoxidised natural rubber toughened poly(lactic acid)/halloysite nanocomposites with high activation energy of water diffusion. J Appl Polym Sci 133(3): 42850.
203. Jiang L, Liu B, Zhang J (2009) Properties of poly(lactic acid)/poly(butylene adipate-co-terephthalate)/nanoparticle ternary composites. Ind Eng Chem Res 48(16): 7594–7602.
204. Ojijo V, Cele H, Ray SS (2011) Morphology and properties of polymer composites based on biodegradable polylactide/poly[(butylene succinate)-co-adipate] blend and nanoclay. Macromol Mater Eng 296(9): 865–877.
205. Ojijo V, Ray SS, Sadiku R (2012) Effect of nanoclay loading on the thermal and mechanical properties of biodegradable polylactide/poly[(butylene succinate)-co-adipate] blend composites. ACS Appl Mater Interf 4(5): 2395–2405.
206. Ojijo V, Ray SS, Sadiku R (2014) Concurrent enhancement of multiple properties in reactively processed nanocomposites of polylactide/poly[(butylene succinate)-co-adipate] blend and organoclay. Macromol Mater Eng 299(5): 596–608.
207. Teo ZX, Chow WS (2016) Impact, thermal and morphological properties of poly(lactic acid)/poly(methyl methacrylate)/Halloysite nanotube nanocomposites. Polym-Plast Technol Eng 55(14):1474–1480.
208. Kelnar I, Kratochvil J, Fortelny I, Kapralkova L, Zhigunov A, Nevoralova M, Kotrisova M, Khunova V (2016) Influence of clay-nanofiller geometry on the structure and properties of poly(lactic acid)/thermoplastic polyurethane nanocomposites. RSC Adv 6(36): 30755–30762.
209. Oliaei E, Kaffashi B (2016) Investigation on the properties of poly(l-lactide)/thermoplastic poly(ester urethane)/halloysite nanotube composites prepared based on prediction of halloysite nanotube location by measuring free surface energies. Polymer 104: 104–114.
210. Jalalvandi E, Majid RA, Ghanbari T, Ilbeygi H (2013) Effects of montmorillonite (MMT) on morphological, tensile, physical barrier properties and biodegradability of polylactic acid/starch/MMT nanocomposites. J Thermoplast Compos Mater 28(4): 496–509.
211. Haroosh HJ, Chaudhary DS, Dong Y (2012) Electrospun PLA/PCL fibers with tubular nanoclay: Morphological and structural analysis. J Appl Polym Sci 124(5): 3930–3939.
212. Bhatia A, Gupta RK, Bhattacharya SN, Choi HJ (2010) Effect of clay on thermal, mechanical and gas barrier properties of biodegradable poly(lactic acid)/poly(butylene succinate) (PLA/PBS) nanocomposites. Int Polym Process 25(1): 5–14.
213. Zhao RX, Torley P, Halley PJ (2008) Emerging biodegradable materials: starch-and protein-based bio-nanocomposites. J Mater Sci 43(9): 3058–3071.
214. Huang J-C, Zhu Z-K, Yin J, Qian X-F, Sun Y-Y (2001). Poly(etherimide)/montmorillonite nanocomposites prepared by melt intercalation: morphology, solvent resistance properties and thermal properties. Polymer: 42(3): 873–877.
215. Barick AK, Tripathy DK (2010) Preparation and characterisation of thermoplastic polyurethane/organoclay nanocomposites by melt intercalation technique: Effect of nanoclay on morphology, mechanical, thermal, and rheological properties. J Appl Polym Sci 117(2): 639–654.

216. Barbosa R, Morais DDS, Nóbrega KC, Araújo EM, Mélo TJA (2012) Influence of processing variables on the mechanical behavior of HDPE/clay nanocomposites. Mater Res 15: 477–482.
217. Puffr R, Špátová JL, Brožek J (2013) Clay mineral/polyamide nanocomposites obtained by in-situ polymerisation or melt intercalation. Appl Clay Sci 83–84: 294–299.
218. Vaia RA, Giannelis EP (1997) Polymer melt intercalation in organically-modified layered silicates: Model predictions and experiment. Macromolecules 30(25): 8000–8009.
219. Le Pluart L, Duchet J, Sautereau H (2005) Epoxy/montmorillonite nanocomposites: influence of organophilic treatment on reactivity, morphology and fracture properties. Polymer 46(26): 12267–12278.
220. Gupta N, Lin TC, Shapiro M (2007) Clay-epoxy nanocomposites: Processing and properties. JOM 59(3): 61–65.
221. Park JY, Davis TB, Sullivan PL (2010). Parametric study on the fabrication of clay-containing thermosetting nanocomposites. J Reinf Plast Compos 29(5): 755–770.
222. Cravotto G, Cintas P (2011) Introduction to sonochemistry: A historical and conceptual overview. In: Chen D, Sharma SK and Mudhoo A (Eds) Handbook on applications of ultrasound: sonochemistry for sustainability. CRC/Taylor and Francis: Boca Raton, 23–40.
223. Dean KM, Sangwan P, Way C, Nikolic MAL (2012). Biodegradability characterisation of polymer nanocomposites. In: Mittal V (Ed) Characterisation techniques for polymer nanocomposites. Wiley: Weinheim, 323–346.
224. Delogu F, Gorrasi G, Sorrentino A (2017) Fabrication of polymer nanocomposites via ball milling: Present status and future perspectives. Prog Mater Sci 86: 75–126.
225. Lonkar PS, Kumar AP, Singh RP (2011) Commercial aspects associated with bio-nanocomposites. In: V. Mittal (Ed) Nanocomposites with biodegradable polymers: Synthesis properties and future perspectives. Oxford University Press: Oxford, 400–421.
226. Koo CM, Ham HT, Choi MH, Kim SO, Chung IJ (2003) Characteristics of polyvinylpyrrolidone-layered silicate nanocomposites prepared by attrition ball milling. Polymer 44(3): 681–689.
227. Ramadan AR, Esawi AMK, Gawad AA (2010) Effect of ball milling on the structure of Na^+-montmorillonite and organo-montmorillonite (Cloisite 30B). Appl Clay Sci 47(3–4): 196–202.
228. Shao W, Wang Q, Wang, F, Chen Y (2006) Polyamide-6/natural clay mineral nanocomposites prepared by solid-state shear milling using pan-mill equipment. J Polym Sci Part B: Polym Phys 44(1): 249–255.
229. Perrin-Sarazin F, Sepehr M, Bouaricha S, Denault J (2009). Potential of ball milling to improve clay dispersion in nanocomposites. Polym Eng Sci 49(4): 651–665.
230. Huang H-C, Huang S-P, Hsieh T-E, Chen C-H (2012) Characterisations of UV-curable montmorillonite/epoxy nanocomposites prepared by a hybrid of chemical dispersion and planetary mechanical milling process. J Appl Polym Sci 123(6): 3199–3207.
231. Ruban YJV, Mon SG, Roy DV (2014) Ball milling: An effective technique for the preparation of exfoliated clay filled unsaturated polyester toughened epoxy nanocomposite. Polym Sci Ser A 56(6): 884–895.
232. Powell CE, Beall GW (2006) Physical properties of polymer/clay nanocomposites. Curr Opin Solid State Mater Sci 10(2): 73–80.
233. Lan T, Pinnavaia TJ (1994) Clay-reinforced epoxy nanocomposites. Chem Mater 6(12): 2216–2219.
234. Jordan J, Jacob KI, Tannenbaum R, Sharaf MA, Jasiuk I (2005) Experimental trends in polymer nanocomposites-A review. Mater Sci Eng A 393(1–2): 1–11.
235. Hussain F, Hojjati M, Okamoto M, Gorga RE (2006) Review article: Polymer-matrix nanocomposites, processing, manufacturing, and application: An overview. J Compos Mater 40(17): 1511–1575.
236. Ishikawa T (2006) Overview of trends in advanced composite research and applications in Japan. Adv Compos Mater 15(1): 3–37.
237. Lu Y-L, Zhang L-Q (2011) Processing methods for the preparation of rubber–clay nanocomposites. In: Galimberti M (Ed) Rubber-clay nanocomposites. Wiley: Weinheim, 145–179.

238. Zare Y (2013) Recent progress on preparation and properties of nanocomposites from recycled polymers: a review. Waste Manage 33(3): 598–604.
239. Aït Hocine N, Seghar S, Hassan H, Azem S (2013) Mechanical performance of thermoset clay nanocomposites. In : Mittal V (Ed) Thermoset nanocomposites. Wiley: Weinheim, 109–128.
240. Albdiry MT, Yousif BF, Ku H, Lau KT (2013) A critical review on the manufacturing processes in relation to the properties of nanoclay/polymer composites. J Compos Mater 47(9):1093–1115.
241. Azeez AA, Rhee KY, Park SJ, Hui D (2013) Epoxy clay nanocomposites – processing, properties and applications: A review. Compos Part B: Eng 45(1): 308–320.
242. Haq M, Burgueño R, Mohanty AK, Misra M (2009) Bio-based unsaturated polyester/layered silicate nanocomposites: Characterisation and thermo-physical properties. Compos Part A: Appl Sci Manuf 40(4): 540–547.
243. Deka H, Karak N (2011) Bio-based hyperbranched polyurethane/clay nanocomposites: adhesive, mechanical, and thermal properties. Polym Adv Technol 22(6): 973–980.
244. Zheng X, Wilkie CA (2003). Nanocomposites based on poly (ϵ-caprolactone) (PCL)/clay hybrid: polystyrene, high impact polystyrene, ABS, polypropylene and polyethylene. Polym Degrad Stab 82(3): 441–450.
245. Chin I-J, Uematsu S (2011) Biodegradation of polymeric systems. In: Mittal V (Ed) Nanocomposites with biodegradable polymers: Synthesis, properties, and future perspectives. Oxford University Press: Oxford, 28–57.
246. Chow WS, Tan SG, Ahmad Z, Chia KH, Lau NS, Sudesh K (2014). Biodegradability of epoxidised soybean oil based thermosets in compost soil environment. J Polym Environ 22(1): 140–147.
247. Tsujimoto T, Uyama H, Kobayashi S (2003) Green nanocomposites from renewable resources: Biodegradable plant oil-silica hybrid coatings. Macromol Rapid Commun 24(12): 711–714.
248. Takahashi T, Hirayama K-I, Teramoto N, Shibata M (2008) Biocomposites composed of epoxidised soybean oil cured with terpene-based acid anhydride and cellulose fibers. J Appl Polym Sci 108(3): 1596–1602.
249. Arora A, Padua GW (2010) Review: nanocomposites in food packaging. J Food Sci 75(1): R43–49.
250. Nielsen LE (1967) Models for the permeability of filled polymer systems. J Macromol Sci Part A: Chem 1(5) 929–942.
251. Fredrickson GH, Bicerano J (1999) Barrier properties of oriented disk composites. J Chem Phys 110(4): 2181–2188.
252. Gusev AA, Lusti HR (2001) Rational design of nanocomposites for barrier applications. Adv Mater 13(21): 1641–1643.
253. Gusev AA (2001) Numerical Identification of the potential of whisker- and platelet-filled polymers. Macromolecules 34(9): 3081–3093.
254. Xu B, Zheng Q, Song Y, Shangguan Y (2006) Calculating barrier properties of polymer/clay nanocomposites: Effects of clay layers. Polymer 47(8): 2904–2910.
255. Choudalakis G, Gotsis AD (2009) Permeability of polymer/clay nanocomposites: A review. Eur Poly J 45(4): 967–984.
256. Ray SS (2013) Clay-containing polymer nanocomposites. Elsevier: Amsterdam, 227–241.
257. Kalendova A, Merinska D, Gerard JF, Slouf M (2013) Polymer/clay nanocomposites and their gas barrier properties. Polym Compos 34(9):1418–1424.
258. Cui Y, Kumar S, Rao Kona B, van Houcke D (2015) Gas barrier properties of polymer/clay nanocomposites. RSC Adv 5(78): 63669–63690.
259. Tan B, Thomas NL (2016) A review of the water barrier properties of polymer/clay and polymer/graphene nanocomposites. J Membr Sci 514: 595–612.
260. Silvestre C, Duraccio D, Cimmino S (2011) Food packaging based on polymer nanomaterials. Prog Polym Sci 36(12):1766–1782.
261. Picard E, Espuche E, Fulchiron R (2011) Effect of an organo-modified montmorillonite on PLA crystallisation and gas barrier properties. Appl Clay Sci 53(1): 58–65.

262. Sabet SS, Katbab AA (2009) Interfacially compatibilised poly(lactic acid) and poly(lactic acid)/polycaprolactone/organoclay nanocomposites with improved biodegradability and barrier properties: Effects of the compatibiliser structural parameters and feeding route. J Appl Polym Sci 111(4): 1954–1963.
263. Gain O, Espuche E, Pollet E, Alexandre M, Dubois P (2005) Gas barrier properties of poly(ε-caprolactone)/clay nanocomposites: Influence of the morphology and polymer/clay interactions. J Polym Sci Part B: Polym Phys 43(2): 205–214.
264. Oguzlu H, Tihminlioglu F (2010).Preparation and barrier properties of chitosan-layered silicate nanocomposite films. Macromol Symp 298(1): 91–98.
265. Rhim J-W, Hong S-I, Ha C-S (2009) Tensile, water vapor barrier and antimicrobial properties of PLA/nanoclay composite films. LWT - Food Sci Technol 42(2): 612–617.
266. Tang X, Alavi S, Herald TJ (2008) Barrier and Mechanical Properties of Starch-Clay Nanocomposite Films. Cereal Chem J 85(3): 433–439.
267. Kim J-K, Hu C, Woo RSC, Sham M-L (2005) Moisture barrier characteristics of organoclay–epoxy nanocomposites. Compos Sci Technol 65(5): 805–813.
268. Alamri H, Low IM (2012). Effect of water absorption on the mechanical properties of nano-filler reinforced epoxy nanocomposites. Mater Des 42: 214–222.
269. Vlasveld DPN, Groenewold J, Bersee HEN, Picken SJ (2005) Moisture absorption in polyamide-6 silicate nanocomposites and its influence on the mechanical properties. Polymer 46(26):12567–12576.

Chapter 2
Material Fabrication and Characterisation Techniques

2.1 Overview

The fabrication of epoxy and bioepoxy/clay nanocomposites was carried out using an in situ polymerisation method. The holistic fabrication of epoxy and bioepoxy/clay nanocomposites regarding this method is illustrated in Fig. 2.1. Bifunctional diglycidyl ether of bisphenol-A (DGEBA) epoxy resin was used as a conventional polymer matrix on the fabrication of epoxy/clay nanocomposites. However, in a bioepoxy/clay nanocomposite system, DGEBA as a conventional epoxy resin is combined with renewable vegetable oils based resin, namely epoxidised soybean oil (ESO), to form DGEBA/ESO blends. Moreover, the polymerisation of DGEBA and ESO/DGEBA blends were prepared based on both amine-based and anhydride curing agents. Meanwhile, four different types of montmorillonite (MMT) nanoclays were utilised as nanofillers. The dispersion process of nanofillers was conducted by using a mechanical shear-mixing technique, and further continued by means of ultrasonication with or without a centrifugal technique. In addition, a vacuuming process was used as the additional step to remove air entrapment produced during the shear mixing process. Various ESO and nanofiller contents were employed to obtain the optimal combination between these two parameters. Furthermore, the parameters of pre-mixing process varied in terms of temperature process, mixing duration, pre-mixing speed, as well as sonication frequency and time.

The morphological structure of bio-based epoxy/clay nanocomposites were determined by X-ray diffraction (XRD) and Fourier transform infrared (FTIR) analyses, which were combined with scanning electron microscopy (SEM), and transmission electron microscopy (TEM). Mechanical tests such as tensile tests (ASTM D638), flexural tests (ASTM D790), impact tests (ASTM 6110) and Durometer hardness test (ASTM D2240) were employed in determining overall mechanical properties of nanocomposites. Meanwhile, differential scanning calorimetry (DSC) was used to investigate the thermal effect on bioepoxy/clay nanocomposites. Nonetheless,

H. Salam and Y. Dong, *Bioepoxy/Clay Nanocomposites*,
https://doi.org/10.1007/978-981-16-7297-2_2

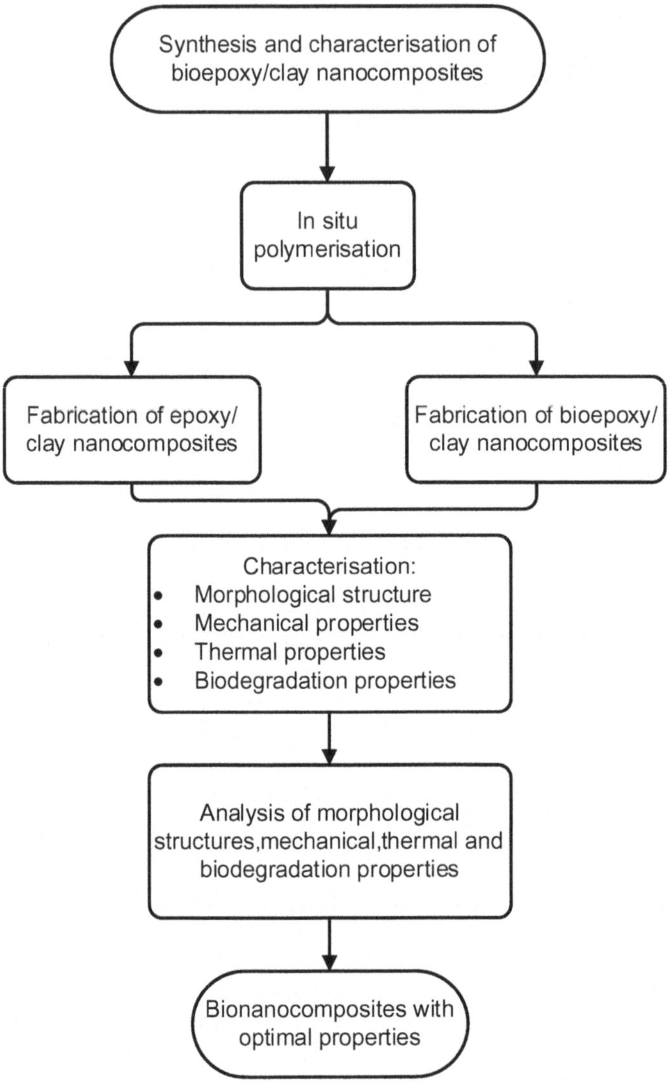

Fig. 2.1 General experimental flowchart

biodegradability tests were conducted to determine the biodegradability of bioepoxy/clay nanocomposites by a composting technique. Whereas water uptake test was employed by the immersion of specimens into the water.

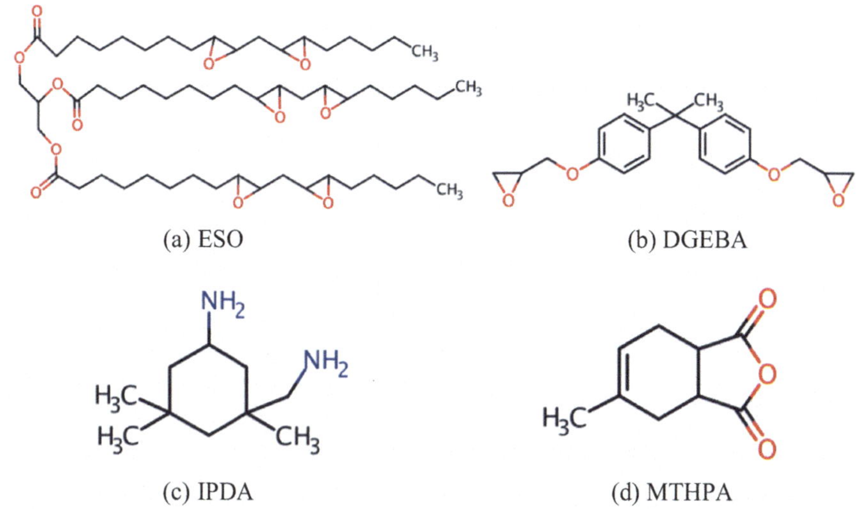

Fig. 2.2 Chemical structure of used materials. **a** ESO, **b** DGEBA, **c** IPDA and **d** MTHPA

2.2 Used Materials

2.2.1 Bioepoxy Blends

Bioepoxy blends were prepared by mixing a partial amount of ESO with conventional epoxy. ESO was supplied and manufacturing by Zhejiang Jiaao Enprotech Stock Co., Ltd., Zhejiang, China. This grade of ESO has an oxirane oxygen content of 6.2% according to the supplier's material data sheet. On the other hand, commercial bifunctional diglycidyl ether of bisphenol-A (DGEBA, L13 Part A) and isophorone diamine (IPDA, L13 Part B) curing agent were purchased from Adhesive Engineering Pty, NSW, Australia. In addition, methyltetrahydrophthalic anhydride (MTHPA) curing agent blended with quaternary ammonium salt latent catalyst (Lindride 6) was supplied by Lindau Chemical, Inc, Columbia, SC, USA. The ratio of epoxy and curing agent was 30 phr for IPDA and 85 phr for MTHPA. DGEBA and ESO were processed with amine-based and anhydride curing agents. Chemical structures of used materials in this study are illustrated in Fig. 2.2.

2.2.2 Clay Nanofillers

Platelet-like layered silicate montmorillonite (MMT) clays were used. Two different types of MMT clay fillers, specified as unmodified clays (Cloisite Na^+) and organomodified clays (i.e., Cloisite 10A, Cloisite 15 and Cloisite 93A) were

Table 2.1 Material properties of DGEBA, ESO and clays

Materials	Properties	Reference			
DGEBA	Epoxy equivalent weight (EEW): 190–195 g/eq	Technical data sheet of L13, Adhesive Engineering Pty			
	Molecular weight: 340.42 g/mol	Technical information of DGEBA resin. National Centre for Biotechnology Information (CID = 2286)			
	Density: 1.17 g/cm^3				
	Viscosity: ~ 11.000 mPa·s				
ESO	Oxirane content: 6.2%	Technical data sheet of ESO, Zhejiang Jiaao Enprotech Stock Co., Ltd.			
	Molecular weight: 975.40 g/mol	Technical information of ESBO. National Centre for Biotechnology Information. (CID=71306824).			
	Density: 0.982 g/cm^3				
	Viscosity: 325 mPa·s				
Clays	E_{MMT} = 178 GPa for single clay platelet	[1]			
MMT clays	Organoclay salt	d_{001} *(nm)*	CEC (meq/100g)	Density (g/cm3)	Moisture (%)
Cloisite Na$^+$		1.17	92.6	2.86	4–9
Cloisite 10A	Benzyl(hydrogenated tallow alkyl) dimethyl, salt	1.90	125	1.90	<3
Cloisite 15	Bis(hydrogenated tallow alkyl) dimethyl, salt	3.63	125	1.66	<3
Cloisite 93A	Trialkyl ammonium, salt	2.36	90	1.88	<2

(Technical data sheet of Cloisite clays are available at https://www.byk.com/en/additives/additives-by-name/cloisite.php)

supplied by BYK-Chemie GmbH, Germany with different types of organic modifiers, interlayer spacings (d_{001}) and cation exchange capacity (CEC) listed in Table 2.1.

2.3 Material Specimen Preparation

In general, the fabrication of conventional epoxy/clay nanocomposites and bioepoxy/clay nanocomposites were prepared via in situ polymerisation. In situ polymerisation was firstly carried out by the direct mixing of nanofillers into conventional epoxy or bioepoxy matrices at different clay contents ranging from 1 to 10 wt%. Three different clay dispersion methods within polymer matrices were employed, namely shear mixing, ultrasonication and centrifugation. Furthermore, the following processes were intended to reduce air entrapment by a vacuuming process with the addition of a certain amount of curing agents for crosslinking reaction. Subsequently, the mixture solution was injected into the testing mould by a syringe to prepare mechanical testing specimens. Finally, the mixtures were cured at different curing steps based on the use of various curing agents.

2.4 Pre-mixing Processes

2.4.1 Shear Mixing

Mechanical shear mixing is a common process for dispersing solid particles into liquid phases. In this process, the fluid shear force produced from an impeller motion can break up clay aggregates into fine clay particles. As previously mentioned, instrument and processing parameters could significantly influence the optimal clay dispersion within polymer matrices.

In this study, mechanical shear mixing was conducted by using an IKA RW20 mechanical mixer coupled with a dispersing blade (Fig. 2.3). Shear processing parameters such as shearing speed, mixing time and temperature varied. Shear speed was in the range from 250 to 500 rpm. Meanwhile, mixing time and temperature were adjusting from 1 to 2 h and 25 to 80 °C, respectively. Those selected shear-mixing parameters are widely used in the fabrication of epoxy/clay nanocomposites as they

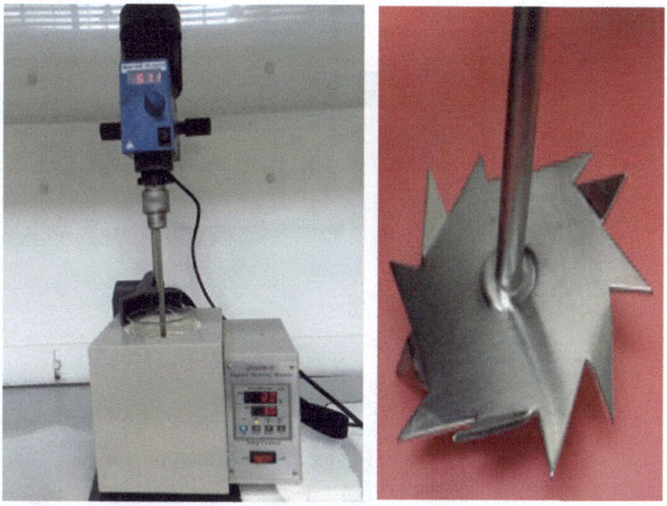

(a) (b)

Fig. 2.3 Shear mixing process: **a** IKA RW20 mechanical mixer with ZNHWII heating mantle and **b** dispersing blade

can achieve better clay dispersion within epoxy matrices and homogenous epoxy/clay mixture [2–8].

2.4.2 *Ultrasonication*

In addition to initial shear mixing process, ultrasonication was also conducted to disperse fine particles into polymer matrices. The sound wave propagation of ultrasonication can produce air-bubble motions that could assist polymer solutions in penetrating into clay interlayer areas, thus resulting in an increase in clay interlayer spacing. In relation to solid–liquid dispersion technique, ultrasonication processing parameters such as ultrasonic power intensity, frequency, temperature and time are well known to play an important role in clay wettability and effective clay dispersion.

An ultrasonication bath ELMA type Ti-H-5 (Fig. 2.4) was employed in this study. Processing parameters, namely sonication frequency and time varied from 25 to 45 kHz and 1 to 3 h, respectively. Sonication frequencies were chosen based on the low and high intensities of ultrasonic settings. Whereas, sonication time was selected according to previous experimental results, which suggested the importance of sufficient time on dispersing clay fillers within matrices [9, 10]. In addition, ultrasonic power intensity and temperature were fixed at 90% and 50 °C accordingly. Those parameters were chosen for optimal clay dispersion to be achieved at the high ultrasonication power while the sonication temperature was maintained at 50 °C.

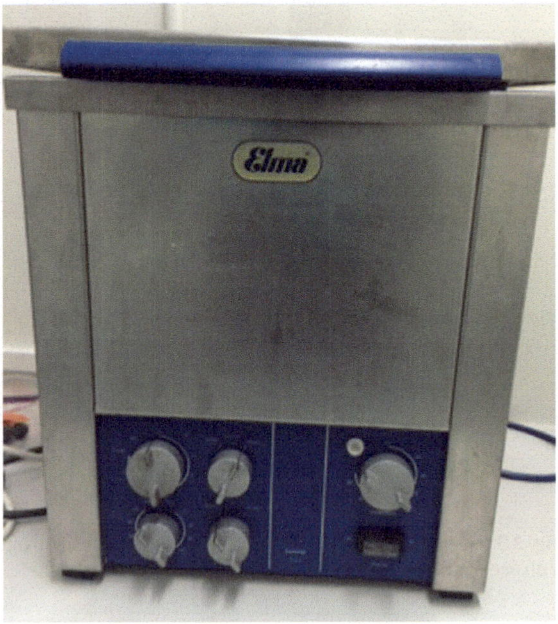

Fig. 2.4 Ultrasonic bath ELMA type Ti-H-5

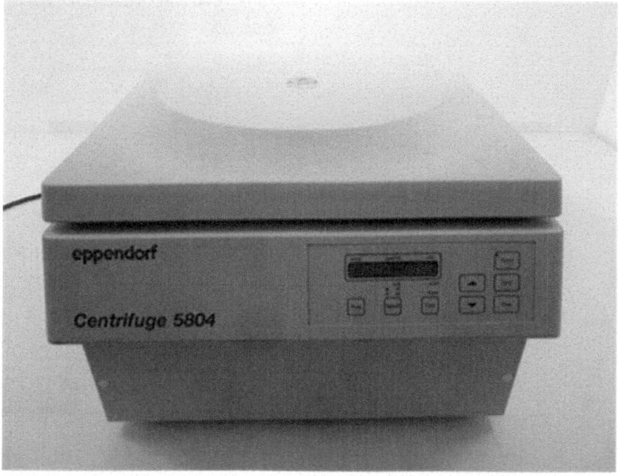

Fig. 2.5 Eppendorf centrifuge 5810R

Furthermore, it is also essential for effective ultrasonic waves to pass through the mixture due to the viscosity reduction [11–13]. Prior to this process, the mixture was poured into a plastic glass cup to prevent any ultrasound barrier arising from thick glass wall of the beaker.

2.4.3 Centrifugation

Alternatively, the centrifugation was used to disperse clay particles into polymer matrices. Generally, such a technique is employed to separate solid phases in the mixture. Meanwhile, the rotational movement at a high speed in the centrifugal tube yields a centrifugal force to facilitate the particle separation. With the particle separation, polymeric molecules can diffuse into clay inter-galleries, thus resulting in an increase in interlayer spacing of clays. Processing parameters such as rotor speed, duration of rotation and processing temperature can highly influence a centrifugal process [14]. In addition, the centrifugation might be potentially used to reduce air entrapment within epoxy/clay mixtures. The centrifugal process was carried out by using an Eppendorf centrifuge 5810R instrument (Fig. 2.5). The mixture was placed in a centrifugal tube at a rotor speed of 4000 rpm for 30 min at 40 °C to remove any entrapped air bubbles.

2.4.4 Vacuuming

Notwithstanding the benefit of using a centrifugal process, the vacuuming process was conducted to remove any air bubble within the mixture. It was performed after the ultrasonication by using a 30 L vacuum oven made by LABEC laboratory equipment

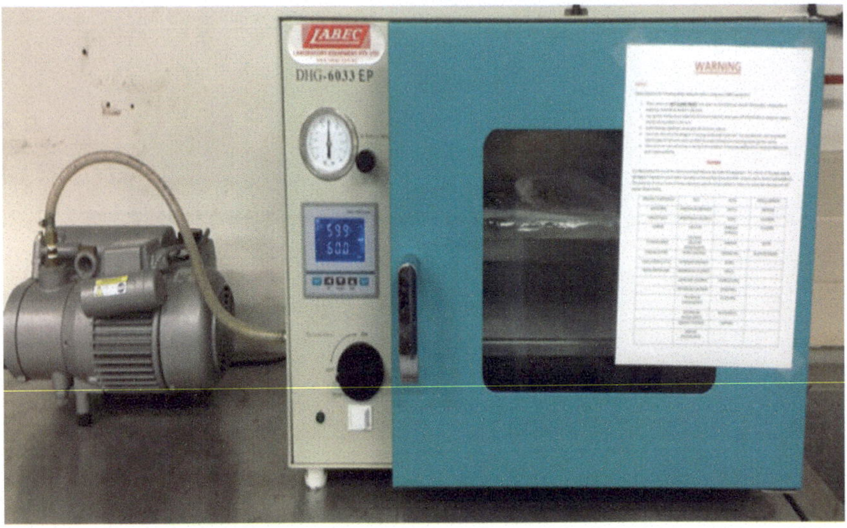

Fig. 2.6 Vacuum oven LABEC DHG-6033EP

(Fig. 2.6). The process was prepared by placing the mixture in a vacuum oven at 50 °C for a few times until it became clear with the reduction of air bubbles. The duration of each vacuuming process was controlled for 2 min.

2.4.5 Casting and Curing

Further processes comprise the mixing of curing agents, solution casting and complete curing. Curing agents were added to the mixture for a good mixing using a glass rod for about 3 min, which was followed by a degassing step on the ultrasonic bath for approximately 5 min. Subsequently, the slurry was injected into mould cavities for mechanical testing specimens using a syringe.

Curing process was carried out in an air-circulating oven with the aid of curing agents (Fig. 2.7). For the mixture cured with IPDA, the process started by curing at 25 °C for 24 h prior to the post-curing at 100 °C for 10 h, and followed by 120 °C for 1 h. Meanwhile, with respect to MTHPA being used as a latent thermal curing agent, the mixture curing was conducted at 80 °C for 24 h, and then followed by the post-curing at 120 °C for 3 h.

Fig. 2.7 Air-circulating oven used in the curing process

2.5 Fabrication of Epoxy/Clay Nanocomposites

2.5.1 Three Combined Pre-mix Processing

The manufacture of epoxy/clay nanocomposites was initialised by pre-heating 100 g DGEBA resin at 80 °C in order to reduce the viscosity of epoxy resin. Desirable amounts of clay particles in range from 3 to 10 wt% were poured slowly and hand mixed into epoxy with a glass rod until all clays were immersed. Clay fillers were pre-dried under vacuum at 80 °C for 24 h. Following this, DGEBA/clay mixture underwent shear mixing with a rotor speed of 510 rpm for 1 h using a shear mixer. Further, the mixture was poured into a plastic sealed bag. This was followed by the ultrasonication at 25 kHz with a sweep mode and 100% power intensity at 50 °C for 45 min to achieve finer clay dispersion. After being poured into a centrifugal tube, the mixture was then placed in a centrifugal machine with a rotor speed of 4000 rpm for 30 min at 40 °C in order to remove entrapped air bubbles. IPDA curing agent was added to epoxy/clay mixture (previously allowed to cool down to ambient temperature) using thorough hand mixing for 5 min. It was followed by the ultrasonication performed at 25 kHz with 100% power intensity in a degassing

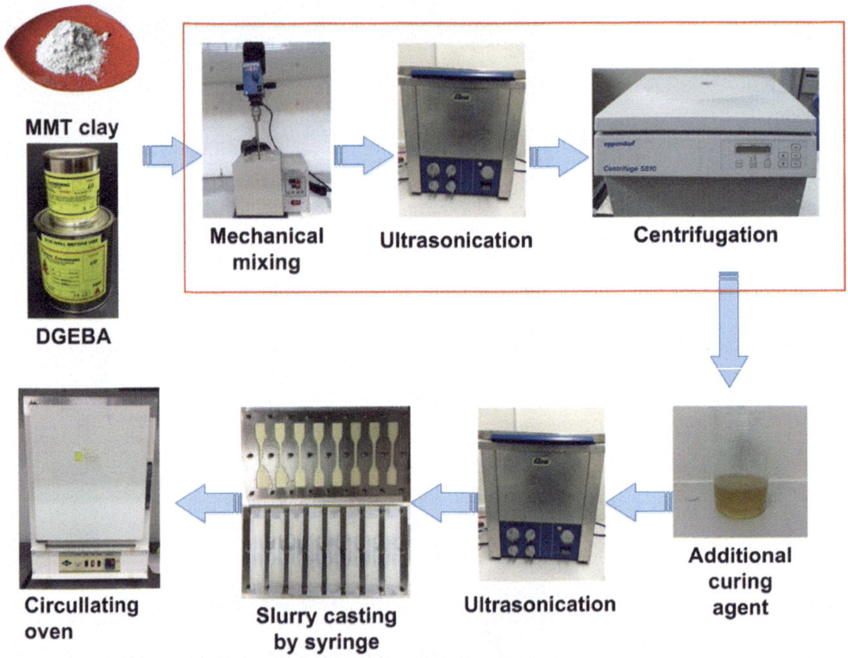

Fig. 2.8 Flowchart of epoxy/clay nanocomposites by three combined premixing processes

mode for 15 min at ambient temperature. The mixed slurry was subsequently poured into mould cavities for the preparation of mechanical testing samples. Finally, all fabricated nanocomposite samples were cured with IPDA. The entire preparation procedure is illustrated in Fig. 2.8.

2.5.2 *Fabrication of Bioepoxy/Clay Nanocomposites*

On the other hand, the fabrication of bioepoxy/clay nanocomposites by applying a robust design of experiments (DoEs) method was conducted to investigate the preferred material formulation and processing parameters. The processing steps to fabricate bioepoxy/clay nanocomposites were illustrated in Fig. 2.9. Bioepoxy blends were prepared by replacing DGEBA epoxy at the ESO contents of 0, 20, 60 and 80 wt% in material formulation. The material mixing was carried out by using a magnetic stirrer at 200 rpm and 25 °C for 15 min. Clay fillers were slowly added at different clay contents from 1, 3, 5 to 8 wt% and mixed with bioepoxy mixture using a glass rod. Prior to this, clay fillers were dried at 80 °C for 24 h. This step was followed by mechanical shear mixing to break up large clay aggregates and then ultrasonication to disperse fine clay particles into bioepoxy matrices. Ultrasonic power intensity and

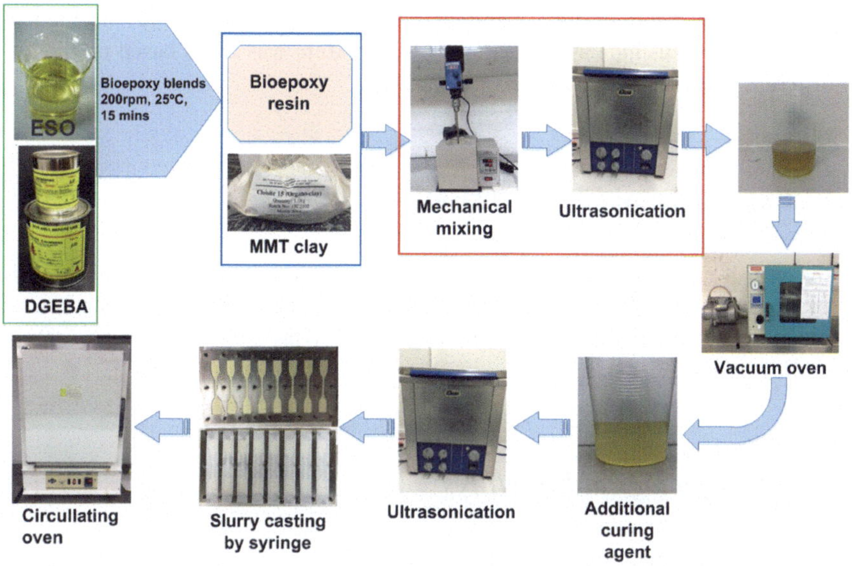

Fig. 2.9 Flowchart for material fabrication of biepoxy/clay nanocomposites

temperature process were adjusted with 90% power intensity at 50 °C, respectively. Material formulation and processing parameters used in this study varied based on the factors and levels of DoEs illustrated in Table 2.2. Subsequently, each mixture was placed in a vacuum oven at 50 °C to remove air bubble entrapment until it became clear. The mixture was then allowed to cool down to ambient temperature. Following this, the curing agent was added to the mixture using a glass rod and stirred for approximately 3 min prior to a degassing step using an ultrasonic bath

Table 2.2 Material formulation and processing parameters for the fabrication of bioepoxy/clay nanocomposites

Factors	Levels			
	1	2	3	4
A: ESO content (wt%)	0	20	60	80
B: Clay type	Cloisite Na^+	Cloisite 10A	Cloisite 15	Cloisite 93A
C: Clay content (wt%)	1	3	5	8
D: Curing agent type	IPDA	MTHPA		
E: Mechanical mixing speed (rpm)	200	500		
F: Mechanical mixing temperature (°C)	25	80		
G: Mechanical mixing time (h)	1	2		
H: Sonication frequency (kHz)	25	45		
I: Sonication time (h)	1	3		

about 5 min. The resulting slurry was then injected into mould cavities of mechanical testing specimens using a syringe. Finally, the mixture was cured based on the use of different types of curing agents.

2.6 Experimental Characterisation

Conventional nanocomposites and bionanocomposites were characterised to evaluate different material features and properties including morphological structures, mechanical and thermal properties, biodegradability, as well as water absorption according to Fig. 2.10.

2.6.1 Morphological Structures

2.6.1.1 X-ray Diffraction (XRD) Analysis

The most popular and simple method to evaluate clay structures in nanocomposites is XRD. In XRD diffractograms, peak characteristics can be observed in terms of the shift in diffraction angle, as well as peak intensity for clay structures. With respect to clay structures, the interlayer spacing between stack-layered clays can increase when the XRD peaks shift to the lower diffraction angles. Furthermore, more periodically oriented clay structures can also be characterised by the high peak intensity, which is influenced by clay content, as well as crystalline structures of polymer/clay nanocomposites. As such, XRD analysis is usually employed to investigate intercalated clay structures in polymer/clay nanocomposites where the insertion of polymeric molecules into clay interlayer areas results in XRD peak shift to lower diffraction angles [15].

The interlayer spacing of clays is determined by Bragg's Law. Bragg's law states that when the condition of the X-ray incident plane onto the surface of a crystal with the initial angle, its reflective angle will diffract into equal value. In addition, the interlayer spacing of the layers is equal to an integer number of the wavelength. In this case, the total length would be equal to a whole number (n) of wavelength (λ) in terms of the order of diffraction. Meanwhile, two waves are subjected to interfaces, which is equal to $2d \sin \theta$ where θ is a measured diffraction angle and d is the interlayer spacing between two clay platelet layers. Accordingly, d for clay platelet layers can be related to 2θ using according to Bragg's law given by

$$n\lambda = 2d \sin\theta \tag{2.1}$$

XRD patterns were obtained using PANalytical EMPYREAN with PIXcel^{3D} X-ray diffractometer (Fig. 2.11) based on CuKα radiation (wavelength $\lambda = 0.1541$ nm)

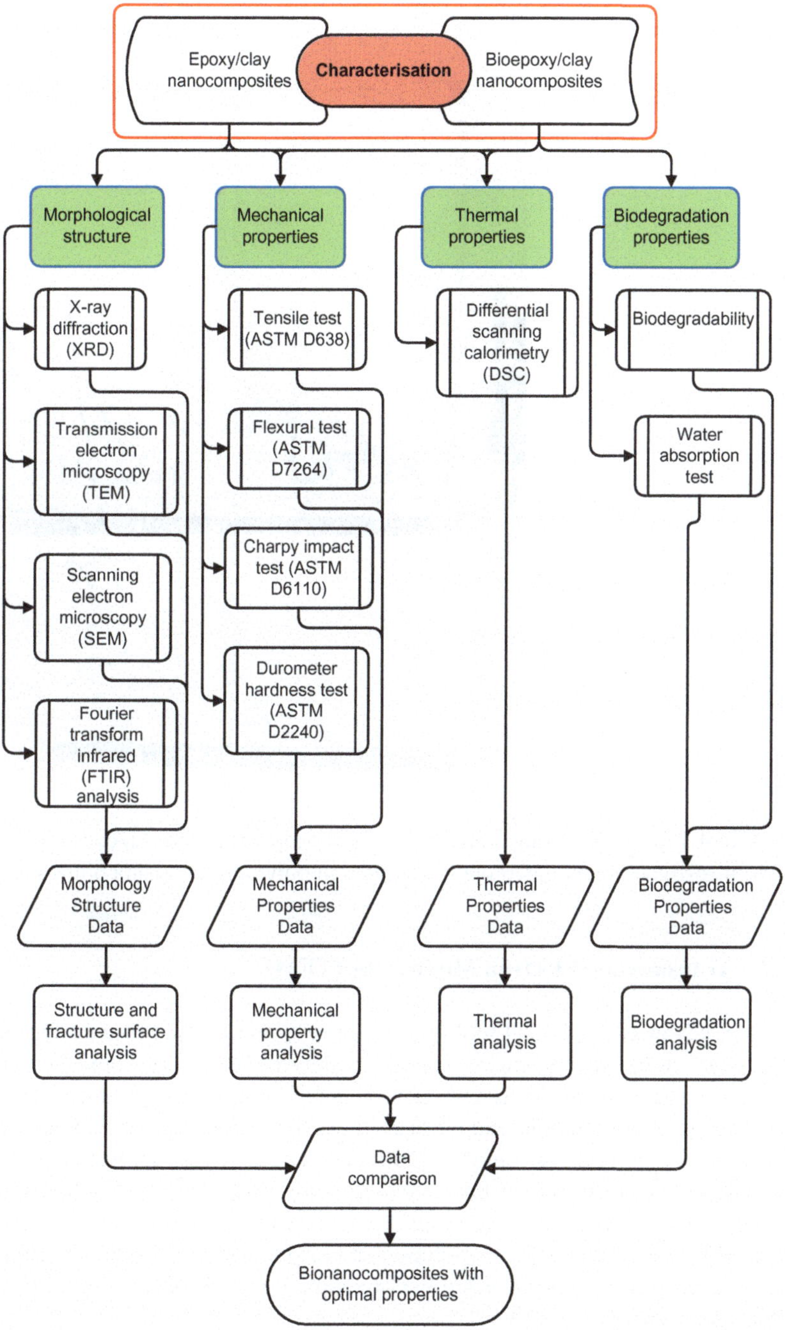

Fig. 2.10 Schematic diagram of material characterisation for conventional epoxy/clay nanocomposites and bioepoxy/clay nanocomposites

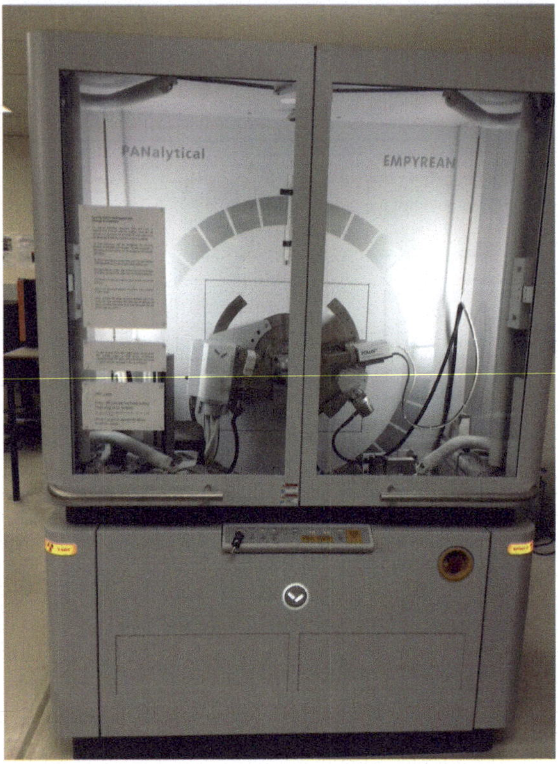

Fig. 2.11 PANalytical EMPYREAN instrument with PIXcel^{3D} X-ray diffractometer

at 40 kV and 40 mA. Scanning step time of 296 s/step with a step size at 0.02° was set up to investigate intercalated clay structures in polymer/clay nanocomposites.

2.6.1.2 Transmission Electron Microscopy (TEM)

TEM is an advanced analytical tool used to observe and analyse actual images of the specimens at the microscaled or nanoscaled levels. This microscopic technique works by using the emitted electron beam through thin specimens to produce morphological structures of materials, graphical features (i.e. texture, shape and size) and composition images of the specimens. TEM technique is usually applied along with XRD analysis to confirm and characterise morphological structures of polymer/clay nanocomposites. TEM analysis could provide images with high magnifications up to 1 nm. Moreover, with the high resolution and magnification, it enables to capture and reveal crystal structures, particle/filler orientations and chemical composition of material specimens.

TEM works with a basic principle similar to optical microscopy. In contract, it applies an electron beam as a light source with the emission from the cathode inside the electron gun located on the top of the column. The emitted electron beam

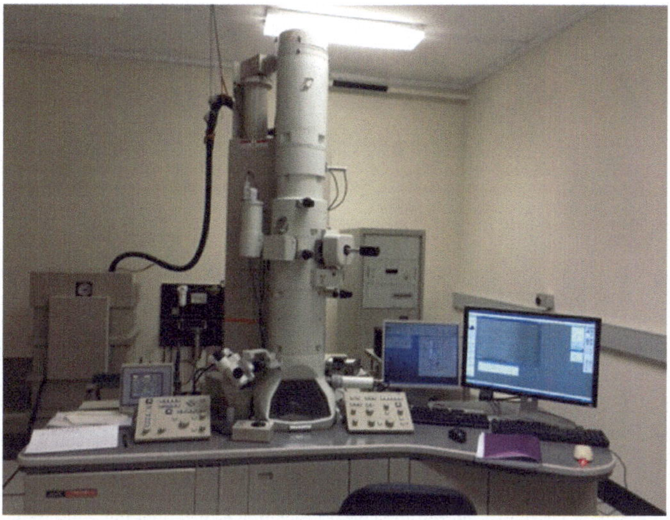

Fig. 2.12 JEOL 2100 TEM instrument

is accelerated by the anode and then rapidly moving to pass through the vacuum column into focusing condenser lenses that are encountered through the specimens. The electron can be absorbed, scattered or pass through the specimens. Various specimen regions would produce different amounts of passed energy, which can eventually influence the contrast of SEM micrographs. The original micrograph is initially in black and white that can be observed in fluorescence screen.

TEM analysis was carried out using a JEOL 2100 TEM instrument (Fig. 2.12) to evaluate the clay dispersion within polymer matrices at an accelerating voltage of 120 kV where LaB6 emitter was used as the electron-beam source. The second largest condenser aperture 2 was inserted to allow a high electron beam to encounter the specimens' surface. In addition, due to the sensitivity of specimens to the beam, the second largest objective aperture was inserted before turning on the electron beam. The images of the specimen were recorded using a Gatan Orius CCD camera with 11 MPixel of graphics resolution. Ultrathin TEM specimens (average thickness: ~100 nm) were sliced at room temperature using a Leica EM UC6 ultramicrotome with a glass knife (Fig. 2.13). The sectioned specimens were collected on 300-mesh copper grids prior to the TEM observation. The sample images were captured at both low and high magnifications in order to observe the clay dispersion status and the formation of clay structures.

2.6.1.3 Scanning Electron Microscopy (SEM)

SEM is another type of microscopic analysis, which is used to capture detailed information of fractured surfaces on bulk materials. The captured images result from the

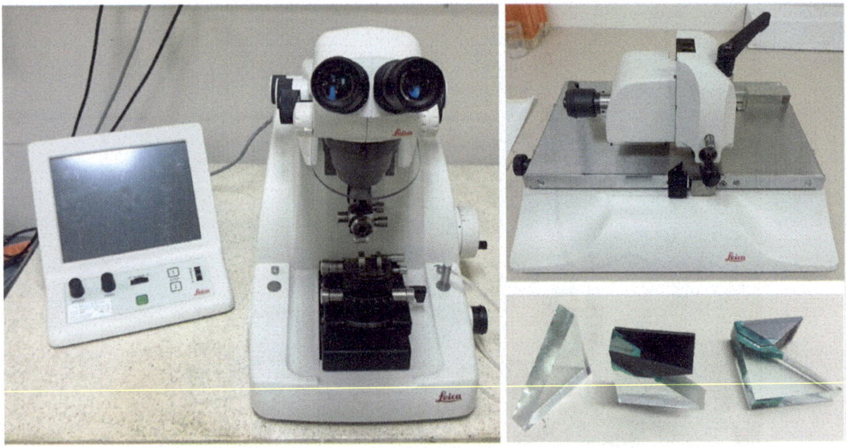

(a) (b)

Fig. 2.13 **a** Leica EM UC6 ultramicrotome and **b** Leica glass knife making equipment

scanning motion of high-energy electron focused beam along the specimen surface areas. Energetic electrons are released from the specimen surface, which generates a variety of signals from electron-specimen interactions. The variety of scattered signals can produce different information on the shape, size, texture and chemical composition of the material specimens.

Tensile fracture surfaces of selected testing specimens were examined by a Zeiss Neon 40EsB FIBSEM (Fig. 2.14a) at an accelerating voltage of 5 kV with the spot size of 30 mm and a working distance of ~5 mm. For bulk materials, the preparation of SEM specimens was initiated by cutting and cleaning fractured surface areas, which were further mounted on a SEM tub (diameter: 12.5 mm) with the aid of double-sided carbon tape (Fig. 2.14b). As for powdered materials, a small amount of material samples was directly deposited and stuck by using a double-sided carbon. All the SEM specimens underwent sputter coating with the platinum layers (layer thickness: 5 nm).

2.6.1.4 Fourier Transform Infrared (FTIR) Analysis

FTIR spectroscopy is well established as an effective method of vibrational spectroscopy, which has been used for decades and considered to be an analytical tool to identify and characterise chemical compositions of material samples. In the FTIR analysis, an infrared spectrum of absorption and emission of solid, liquid or gas samples can be determined. In this study, FTIR spectra were recorded in the wavenumber range of 4000–400 cm^{-1}. Perkin-Elmer FTIR-spectrum 100 spectrophotometer (Fig. 2.15) was employed to record chemical compositions of bioepoxy/clay

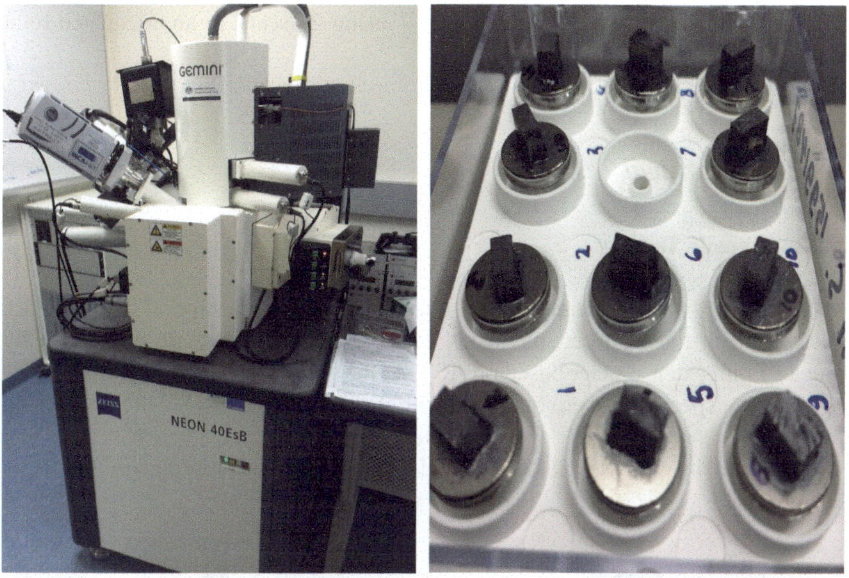

(a) (b)

Fig. 2.14 SEM analysis: **a** Zeiss Neon 40EsB FIBSEM and **b** sputter-coated SEM specimens

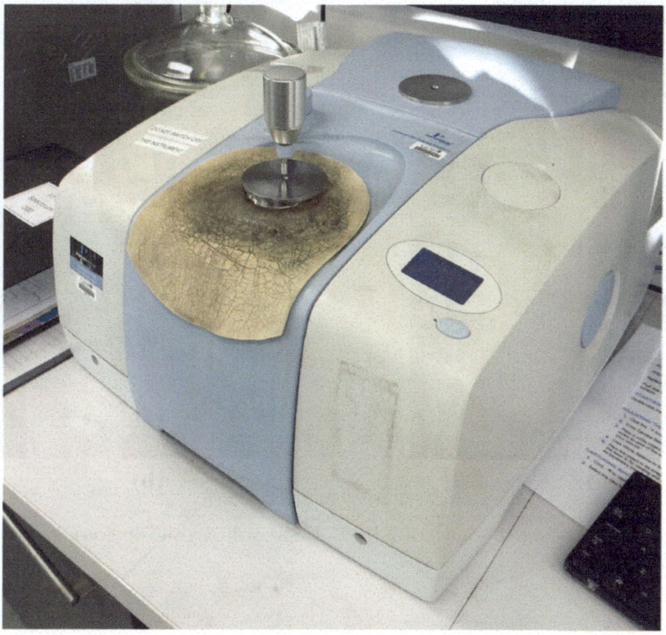

Fig. 2.15 Perkin-Elmer FTIR-spectrum 100 spectrophotometer

nanocomposites with a resolution of 4 cm^{-1} using 20 scans in an attenuated total reflectance (ATR) correction mode.

2.6.2 *Mechanical Testing*

2.6.2.1 Tensile Testing

Tensile testing was performed on a universal testing machine (Lloyds EZ50, Fig. 2.16) with the load cell of 50 kN at the crosshead speed of 5 mm/min by using a 25 mm extensometer based on ASTM D638. The dimensions of "dog-bone" tensile testing specimens were in accord with ASTM D638 type IV. Ultimate tensile strength and tensile modulus were reported from the average data based on five testing specimens with associated standard deviations for test reproducibility.

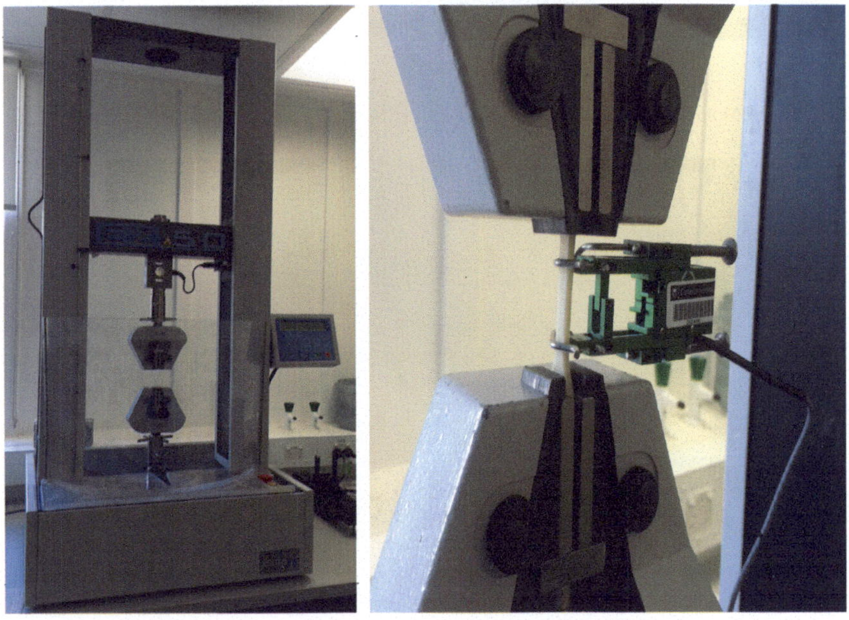

(a) **(b)**

Fig. 2.16 Universal testing machine: **a** Lloyds EZ50 with tensile test grips and **b** clip-on extensometer

2.6.3 *Flexural Testing*

Flexural tests were carried out based on a three-point bending rig on a universal testing machine (Lloyds EZ50, Fig. 2.17) according to ASTM D790. The recommended specimens are in size of 127 × 12.7 × 3.2 mm. The support span length is 16 times of specimen thickness. Accordingly, the support span L is 51.2 mm in length. However, the rate of crosshead motion R (mm/min) is calculated according to the following equation

$$R = \frac{ZL^2}{6d} \tag{2.2}$$

where d is the depth of beam (mm) and Z is the rate of straining of the outer fibre (mm/mm/min) equal to 0.01.

Additionally, according to ASTM D790 in relation to the testing procedure, flexural tests should be completed at break or at a maximum strain of 0.05 mm/mm. The deflection at the maximum strain can be calculated as follows

$$D = \frac{rL^2}{6d} \tag{2.3}$$

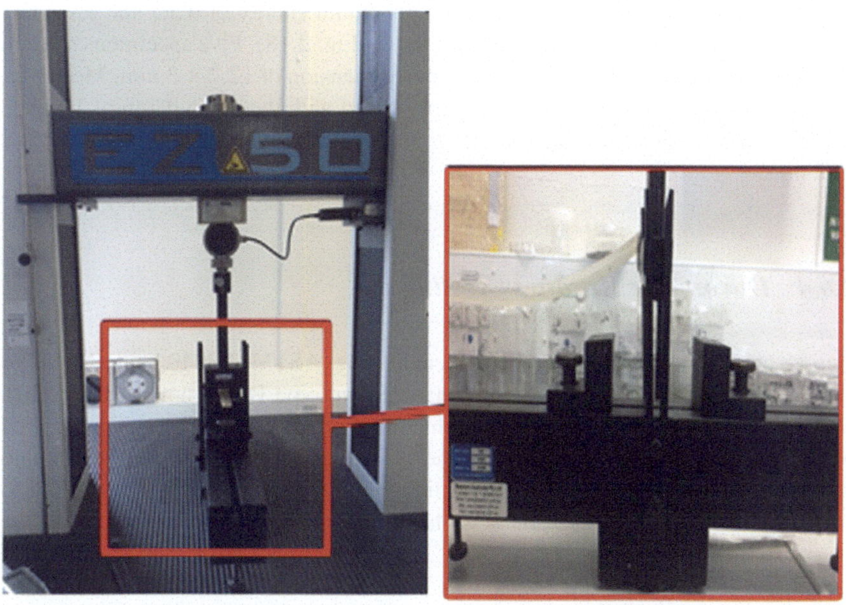

Fig. 2.17 Three-point bending rig on a universal testing machine Lloyds EZ50

where r (maximum strain) is equal to 0.05 mm/. As such, the maximum deflection is equal to 6.28 mm accordingly.

On the other hand, flexural strength (σ_f), flexural strain (ε_f) and flexural modulus of elasticity (E_B) are calculated according to Eqs. 2.4–2.6 below, respectively

$$\sigma_f = \frac{3PL}{2bd^2} \tag{2.4}$$

$$\varepsilon_f = \frac{6Dd}{L^2} \tag{2.5}$$

$$E_B = \frac{L^3 m}{4bd^3} \tag{2.6}$$

in which P is the load applied at the given point on the load–deflection curve (N), b is the width of beam tested (mm) and m is the slope of the tangent to the initial straight-line position of the load deflection curve (N/mm).

2.6.3.1 Charpy Impact Testing

Charpy impact tests with the formation of V notch were carried out based on ASTM D6110 with a pendulum arm of 225 mm, pendulum energy of 0.5 J and a span of 101.6 mm using a Zwick D-7900 impact tester (Fig. 2.18). Five specimens of each material batch were tested with the specimen dimension of 12.7 × 3 mm. Moreover, impact specimens were also notched before tested with the notch depth of 10.16 ± 0.05 mm, the notch radius angle of 45 ± 1° of and the radius curvature of 0.25 ± 0.05 mm.

2.6.4 Durometer Hardness Testing

Durometer hardness tests were conducted using ASTM D2240 with Mitutoyo analogue Durometer 811–337-01 hardmatic Shore type D instrument (Fig. 2.19). The testing specimens were 6 mm in thickness or were comprised of multiplied pieces to obtain necessary thickness. The surface of specimens should be flat and parallel along the testing area. At least five measurement points were analysed to evaluate the surface characteristic of each specimen with at least 6 mm apart.

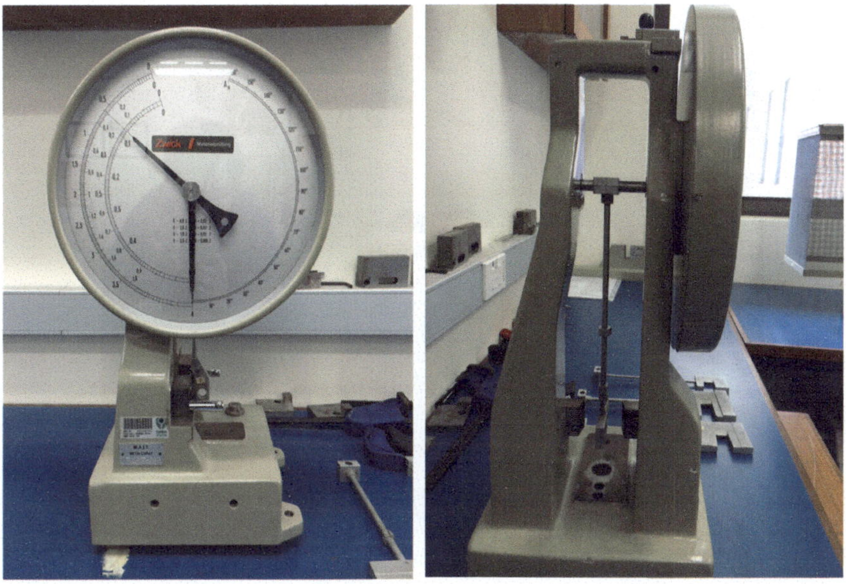

(a) (b)

Fig. 2.18 Zwick D-7900 impact tester: **a** front view and **b** side view

Fig. 2.19 Mitutoyo analogue Durometer 811–337-01 hardmatic Shore type D

2.6.5 Differential Scanning Calorimetry (DSC)

DSC is a thermal technique to investigate thermodynamic processes and kinetic phenomena of polymeric materials. DSC is used for polymeric materials including the observation of glass transition temperature (T_g), melting temperature (T_m), curing phenomenon, degree of crystallinity, compatibility of multiphase and polymerisation process [16]. In this study, the T_g of polymers and nanocomposite materials was identified using a midpoint method provided by Pyris™ data analysis software. It

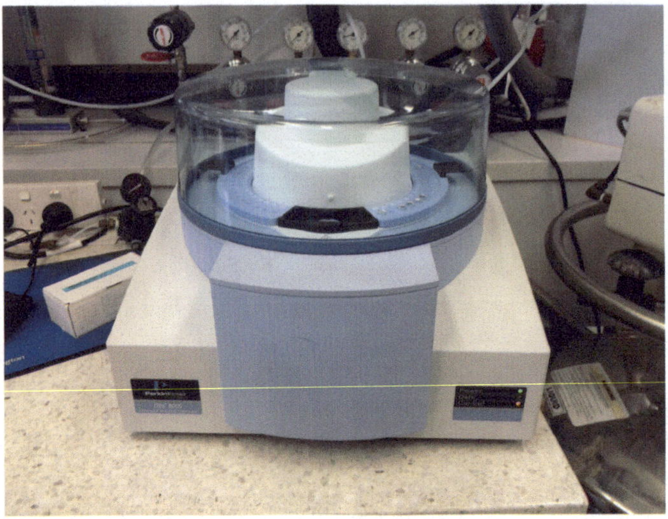

Fig. 2.20 PerkinElmer DSC6000 thermal analyser

was determined as the inflection point of heat-flow temperature curve in the glassy and rubbery regions [17].

In this work, DSC analysis of clay bionanocomposites was performed on a PerkinElmer DSC6000 thermal analyser (Fig. 2.20). Small amounts of cut specimens in a mass range of 5–10 mg were heated from −80 to 300 °C at the heating rate of 10 °C/min under a nitrogen flow of 20 mL/min.

2.6.6 Composting Tests

The biodegradability of polymer/clay bionanocomposites was determined by exposing nanocomposite specimens (30 × 12.7 × 3.2 mm) into active compost for about 180 days, as illustrated in Fig. 2.21. The environmental conditions of composting were maintained every day at an average temperature of 25 ± 2 °C with the relative humidity of 50 ± 5% to keep a pH value of ~7. In addition, the soil compost came from organic fraction, and was sieved on a screen with a mesh size being less than 10 mm. Meanwhile, the ratio of dry soil compost to samples was 6:1 in the total volume of the testing vessel. A total of 15 samples for each material batch were prepared and buried with a distance of 3 cm from each other. Three samples were taken out every 15 days from 0 to 180 days after the exposure in the vessel. The physical appearance and features of specimens (i.e. colour, shape, size or any visible change on the specimen surfaces) were recorded.

In addition, SEM was also employed to look into the details of physical surface changes (i.e. crack, hole or biofilm formation) and microorganism growth during

Fig. 2.21 Composting vessels

composting tests. Specimens from three different time periods (i.e. 0, 3 and 6 months) in composting tests were selected to assess surface changes. The specimens were coated with the platinum layers (thickness: 3 nm). SEM analysis was examined by a Zeiss Neon 40EsB focused ion beam SEM (FIBSEM) instrument at an accelerating voltage of 5 kV, along with a spot size of 30 mm at the working distance of ~5 mm.

On the other hand, the weight loss assessment of the specimens was measured at specific time intervals of 15, 30, 45, 93, 116, 138 and 180 days in composting tests. The specimens were removed from composting vessels and washed with distilled water. This step was followed by drying to a constant weight at 60 °C in the vacuum oven for about 24 h. The percentage of weight loss in terms of biodegradability was calculated below

$$\%W_{loss} = \frac{W_{to} - W_{ts}}{W_{to}} \times 100 \quad (2.7)$$

where $\%W_{loss}$ is the percentage of weight loss, W_{to} and W_{ts} are the weights of specimen at time $t = 0$ (i.e. before exposure) and at a given sampling time, respectively [18].

2.6.7 Water Absorption

Water absorption tests have two functions including an optimal quantity of water absorption by a material and control testing for material uniformity. The moisture

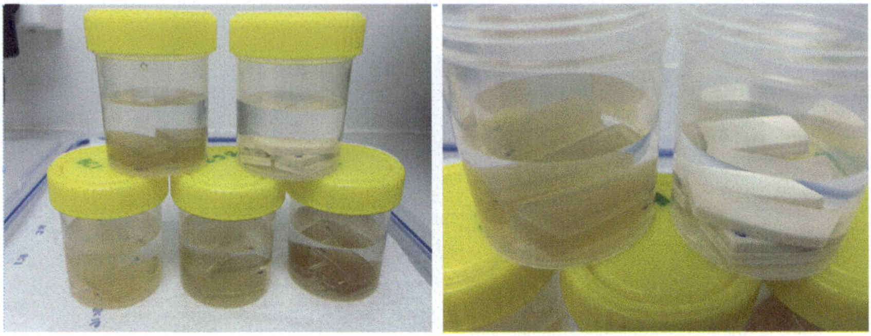

(a) (b)

Fig. 2.22 Water absorption tests: **a** stacked sample containers and **b** immersed material samples

content of polymers is related to electrical insulation resistance, dielectric loose, mechanical strength, appearance and dimension. Additionally, the changes of moisture features can be associated with the specimen shape and intrinsic properties of materials.

Water absorption tests were carried out by the immersion of specimens in deionised water shown in Fig. 2.22. Before that, the specimens were dried in a vacuum oven at 50 °C for approximately 24 h, which were further cooled in a desiccator and then directly weighed. Afterwards, the specimens were substantially immersed in deionised water at ambient temperature. The weight increase was recorded periodically in the time interval up to the average of increase in weight being less than 1%. The percentage of increase in weight during immersion time ($\%W_{gain}$) is given by

$$\%W_{gain} = \frac{W_{ts} - W_{to}}{W_{to}} \times 100\% \tag{2.8}$$

2.7 Summary

Overall, different sophisticated characterisation techniques were implemented to investigate the multifaceted properties of prepared conventional epoxy/clay nanocomposites and bioepoxy/clay nanocomposites with the detailed procedures and protocols in order to establish their processing-structure–property relationship in the forthcoming chapters.

References

1. Fornes TD, Paul DR (2003) Modeling properties of nylon 6/clay nanocomposites using composite theories. Polymer 44(17): 4993–5013.
2. Shi Y, Kanny K, Jawahar P (2009) Hybrid nanocomposites: Processing and properties. Adv Compos Mater 18(4):365–379.
3. Alateyah AI, Dhakal HN, Zhang ZY (2013) Processing, properties and applications of polymer nanocomposites based on layer silicates: A review. Adv Polym Technol 32(4):21368.
4. Atiemo-Obeng VA, Penney W, Armenante PM (2004) Solid-liquid Mixing. In: Paul EL, Atiemo-Obeng VA, Kresta SM (Eds), Handbook of industrial mixing: Science and practice. John Wiley and Sons: Hoboken. 543–584.
5. Park JY, Davis TB, Sullivan PL (2010). Parametric study on the fabrication of clay-containing thermosetting nanocomposites. J Reinf Plast Compos 29(5): 755–770.
6. Gupta N, Lin TC, Shapiro M (2007) Clay-epoxy nanocomposites: Processing and properties. JOM 59(3): 61–65.
7. Dong Y, Chaudhary D, Ploumis C, Lau, K-T (2011) Correlation of mechanical performance and morphological structures of epoxy micro/nanoparticulate composites. Compos Part A Appl Sci Manuf 42(10): 1483–1492.
8. Salam H, Dong Y, Davies IJ. (2015) Development of Bio-based Polymer/Clay Nanocomposites: A Critical Review. In: Dong Y, Umer R, Lau A K-T (Eds) Fillers and Reinforcements for Advanced Nanocomposites. Woodhead Publishing: Cambridge, 101–132.
9. Lam C-K, Lau K-T, Cheung H-Y, Ling H-Y (2005) Effect of ultrasound sonication in nanoclay clusters of nanoclay/epoxy composites. Mater Lett 59(11): 1369–1372.
10. Liu, Z. S., Erhan, S. Z., and Xu, J. (2005). Preparation, characterisation and mechanical properties of epoxidised soybean oil/clay nanocomposites. *Polymer, 46*(23), 10119–10127.
11. Wang R, Schuman T, Vuppalapati RR, Chandrashekhara K (2014) Fabrication of bio-based epoxy-clay nanocomposites. Green Chem 16(4): 1871–1882.
12. Bensadoun F, Kchit N, Billotte C, Bickerton S, Trochu F, Ruiz E (2011) A study of nanoclay reinforcement of biocomposites made by liquid composite molding. Inter J Polym Sci 2011(10): 964193.
13. Kaboorani A, Riedl B, Blanchet P (2013) Ultrasonication technique: A method for dispersing nanoclay in wood adhesives. J Nanomater 2013: 341897.
14. Ford, TC, Graham, JM (1991) Introduction to Centrifugation. New York: Garland Science.
15. Alexandre M, Dubois P (2000) Polymer-layered silicate nanocomposites: Preparation, properties and uses of a new class of materials. Mater Sci Eng R Rep 28(1–2): 1–63.
16. Miracle DB, Donaldson SL (2001) Thermal Analysis. ASM Handbook, Volume 21—Composites. ASM International: Ohio, 973–976.
17. Mutlur S (2004) Thermal analysis of composites using DSC. In: Kessler MR (Ed) Advanced Topic in Characterisation of Composites. Trafford Publishing: Bloomington, 11-33.
18. Dean KM, Sangwan P, Way C, Nikolic MAL (2012) Biodegradability Characterisation of Polymer Nanocomposites. In: Mittal V (Ed) Characterisation Techniques for Polymer Nanocomposites. Wiley: Weinheim, 323-346.

Chapter 3
The Effects of Material Formulation and Manufacturing Process on Mechanical and Thermal Properties of Conventional Epoxy/Clay Nanocomposites

3.1 Morphological Structures of Epoxy/Clay Nanocomposites

3.1.1 X-ray Diffraction (XRD) Analysis

XRD patterns of platelet-like clays and their corresponding nanocomposites were presented in Fig. 3.1. Calculated *d*-spacing values of as-received organolcays and epoxy nanocomposites with different clay types and contents were given in Table 3.1. Overall, the addition of organoclays into epoxy matrices was found to increase *d*-spacing values for all three clay types, namely Cloisite Na^+, Cloisite 10A and Cloisite 93A. However, the *d*-spacing value of epoxy/Cloisite 15 clay nanocomposites remained at a similar level to that of as-received Cloisite 15 clays. It was clearly shown that epoxy nanocomposites reinforced with Cloisite Na^+, Cloisite 10A, Cloisite 15 and Cloisite 93A clays exhibited the increases in interlayer spacing values by approximately 33, 44, 6 and 40%, respectively. This implied that clay intercalation at different levels took place during the manufacture of epoxy/clay nanocomposites owing to the effect of cation exchange capacity (CEC) for different clay types and curing agents [2–4].

More interestingly, the *d*-spacing values of epoxy nanocomposites reinforced with Cloisite 15, Cloisite 10A and Cloisite 93A clays had similar values close to 3.5 nm. It is worth noting that bisphenol A diglycidyl ether (DGEBA) molecules can only be swollen into clay interlayer areas in order to increase *d*-spacing values. Molecular structures and chemical compositions of epoxy resin can play an important role in final morphological structures of resulting nanocomposites, as previously mentioned by Becker et al. [5]. Furthermore, the mobility of epoxy molecules into clay galleries, as well as the inter- and extra-gallery reaction of epoxy should also be taken into consideration.

Alkylammonium modified clay chains are able to occupy the gaps in interlayer areas, which decreases the diffusion of epoxy molecules, thus resulting in a limited

H. Salam and Y. Dong, *Bioepoxy/Clay Nanocomposites*,
https://doi.org/10.1007/978-981-16-7297-2_3

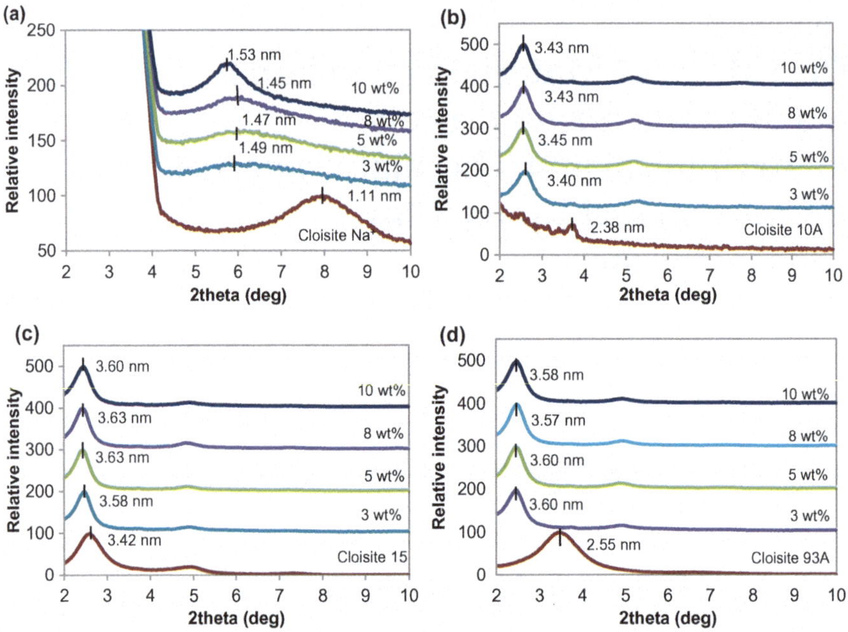

Fig. 3.1 XRD patterns of epoxy/clay nanocomposites based on different clay types and contents: **a** Cloisite Na^+, **b** Cloisite 10A, **c** Cloisite 15 and **d** Cloisite 93A [1]

increase in *d*-spacing values [4]. On the other hand, hydrophilic properties of pristine MMT clays (i.e. Cloisite Na^+) may contribute to a relatively small increase in *d*-spacing up to only 1.4 nm for nanocomposites. Conversely, organoclays modified by organophilic and alkylammonium cations tend to possess essentially hydrophobic clay surfaces with expanded interlayer areas, thus further assisting in the diffusion of epoxy molecules [6–8].

3.1.2 Transmission Electron Microscopic (TEM) Observation

As a supplementary technique to XRD analysis for the investigation of clay dispersion, TEM observation was carried out on epoxy/Cloisite 93A clay nanocomposites at various clay contents, as illustrated in Fig. 3.2. It is evidently shown from these TEM micrographs that combining three different pre-mixing processes tend to produce non-uniform clay dispersion, as evidenced by the presence of clay aggregates with skewed nanoclay layers. The addition of 3 wt% clays resulted in a mix of morphological structures, comprising highly agglomerated and partially intercalated/exfoliated clay structures (Fig. 3.2a). Increasing the clay content demonstrated a high tendency for the provision of intercalation structures rather than exfoliated counterparts, as depicted in Figs. 3.2b, d, f and h. It is clearly shown that the presence of intercalated

Table 3.1 Summary of *d*-spacing values for as-received clays and epoxy/clay nanocomposites [1]

No	Material sample	XRD Peak		
		2θ	Interlayer spacing d_{001} (nm)	% increase of interlayer spacing
1	Cloisite Na^{+}	7.94	1.11	
2	Cloisite Na^{+}-3wt%	5.94	1.49	33.54
3	Cloisite Na^{+}-5wt%	5.99	1.47	32.37
4	Cloisite Na^{+}-8wt%	6.10	1.45	30.09
5	Cloisite Na^{+}-10wt%	5.76	1.53	37.80
6	Cloisite 10A	3.71	2.38	
7	Cloisite 10A-3wt%	2.60	3.40	42.60
8	Cloisite10A-5wt%	2.56	3.45	44.69
9	Cloisite10A-8wt%	2.57	3.43	44.10
10	Cloisite10A-10wt%	2.57	3.43	44.15
11	Cloisite 15	2.58	3.42	
12	Cloisite15-3wt%	2.47	3.58	4.70
13	Cloisite15-5wt%	2.43	3.63	6.19
14	Cloisite15-8wt%	2.44	3.63	6.04
15	Cloisite15-10wt%	2.45	3.60	5.31
16	Cloisite 93A	3.47	2.55	
17	Cloisite 93A-3wt%	2.45	3.60	41.31
18	Cloisite 93A-5wt%	2.46	3.60	41.06
19	Cloisite 93A-8wt%	2.48	3.57	39.89
20	Cloisite 93A-10wt%	2.47	3.58	40.51

structures with an interlayer spacing shifted from 2.55 nm to approximately 3.6 nm (Table 3.1). Moreover, exfoliated clay structures with single clay platelets were only evidently observed in Fig. 3.2b, d and f.

On the other hand, increasing the clay content from 3 to 10 wt% caused silicate layers to clump together, as indicated by the presence of clay aggregates with typical particle sizes ranging from 1 to 3 μm. The abovementioned fact implied that the combination of three different pre-mixing methods with those processing parameters could not produce desirable epoxy/clay nanocomposites with better clay wettability. The processing parameters of mechanical mixing and ultrasonication selected in this study may lack the effective manufacturing of nanocomposites to break up clay aggregates and disperse fine clay particles uniformly in epoxy matrices.

The other plausible reason for clay agglomeration can be attributed to applying centrifugal forces in the third stage of pre-mixing processes, which allows the silicate layers of clay particles to stick together, thus leading to the reduced functionality of clays as effective nanofillers. Such a phenomenon is in good accordance with previous results obtained from Saber-Samandari et al. [9] by investigating the

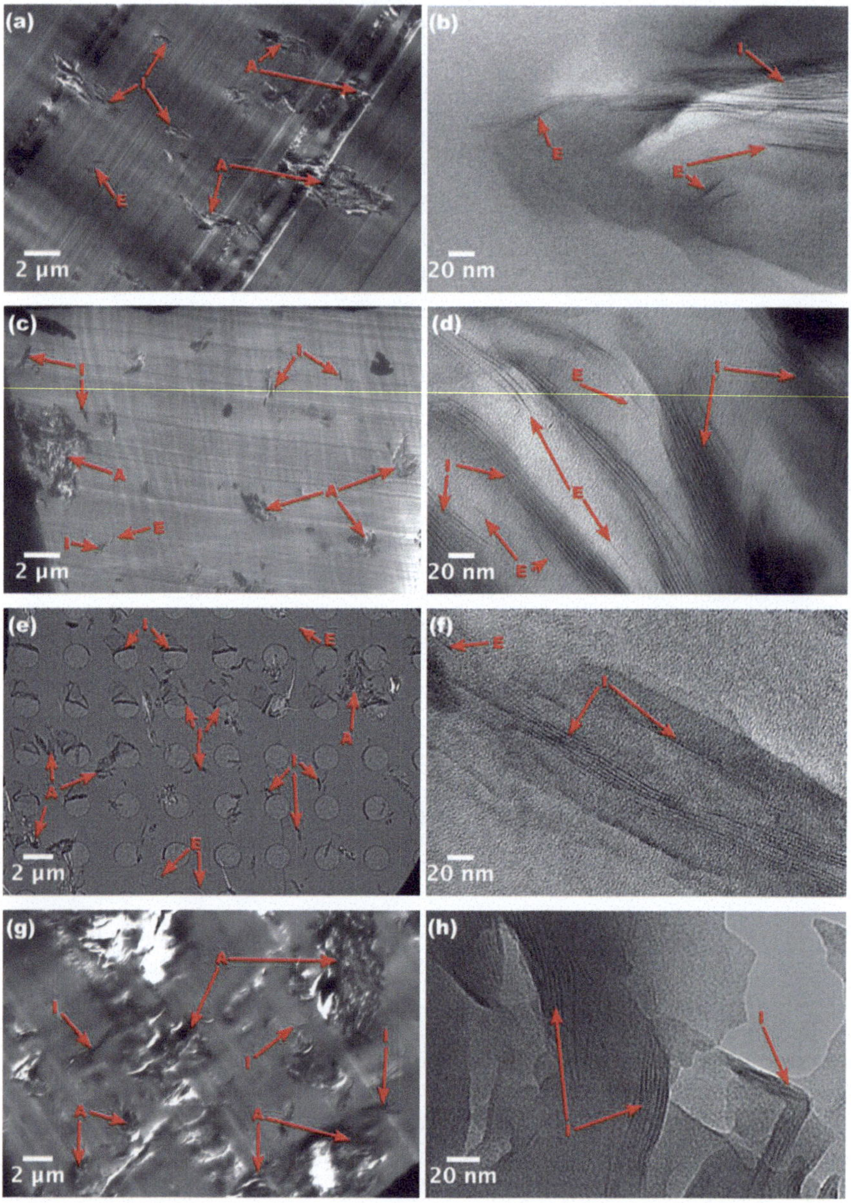

Fig. 3.2 TEM micrographs of epoxy/Cloisite 93A clay nanocomposites with different clay contents: **a** 3 wt% (1200 ×), **b** 3 wt% (80,000 ×), **c** 5 wt% (1200 ×), **d** 5 wt% (80,000 ×), **e** 8 wt% (1200 ×), **f** 8 wt% (80,000 ×), **g** 10wt% (1200 ×) and **h** 10 wt% (80,000×). The letters A, I and E represent agglomerated, intercalated and exfoliated clay structures, respectively and the values in the parentheses are the TEM magnification levels [1]

effect of centrifugal mixing at high rotor speeds on clay agglomeration. Consequently, optimum processing condition at different clay types and contents, as well as compatibility of physical and chemical properties of used coupling agents are the key to successfully manufacture epoxy/clay nanocomposites with better clay wettability and excellent mechanical properties.

3.1.3 Scanning Electron Microscopic (SEM) Analysis

The fracture surfaces of neat epoxy and epoxy/clay nanocomposites investigated by SEM analysis are presented in Fig. 3.3. Neat epoxy specimen exhibited typical brittle fracture surfaces with so-called "stream" marks adjacent to initial cracks (Fig. 3.3a). On the other hand, epoxy/clay nanocomposites possess significantly different morphological structures with a clear sign of much rougher fracture surfaces (Figs. 3.3b–i). Moreover, it was revealed that increasing the clay content in nanocomposites further promoted the surface roughness level, signifying more brittle feature with embedded rigid clay fillers.

On the other hand, the presence of clay aggregates with the size of approximately 2 to 5 μm, as seen in Figs. 3.3b–i, indicated the pull-out of clay particles in the fracture mechanism owing to weak interfacial bonding between clay aggregates and epoxy matrices. Such an observation was believed to further deteriorate mechanical properties of epoxy/clay nanocomposites. Moreover, increasing the clay content up to 10 wt% induced severe clay clumps and agglomeration according to previous TEM results illustrated in Fig. 3.2. In general, the nanofiller content beyond 5 wt% may make it difficult to produce highly dispersed nanofillers in a nanocomposite system. The other possible reason for clay agglomeration can be due to an inappropriate selection of processing parameters to produce insufficient dissipation energy for particle separation. In order to achieve good clay wettability, it is essential to use high shear forces to overcome Van der Waals interactions among clay layers so that epoxy molecules can be easily diffused into clay interlayer areas and achieve the expansion of interlayer spacing, further resulting in well dispersed clay fillers within polymer matrices [10].

Further, Figs. 3.4a–h demonstrated a large number of microvoids and microcracks for epoxy nanocomposites reinforced with organomodified and pristine clays, namely Cloisite 10A, Cloisite 15, Cloisite 93A and Cloisite Na^+ clays. The crack propagation was initiated close to microvoids and clay aggregates due to weak epoxy network junctions around void edges. Therefore, the existence of microvoids can be attributed to several effects that may have taken place in three aforementioned different pre-mixing processes. During the shear mixing process, foamy and viscous mixed materials can be produced, which are hard to be reduced in the following ultrasonication and centrifugal processes. Inappropriate mixing with a fast rotor speed can yield air entrapment or foamy mixture, built-up heat and poor clay dispersion, illustrated from SEM results in Figs. 3.3 and 3.4. Despite the positive effect of a centrifugal technique on bubble reduction [9], it may also break down large bubbles

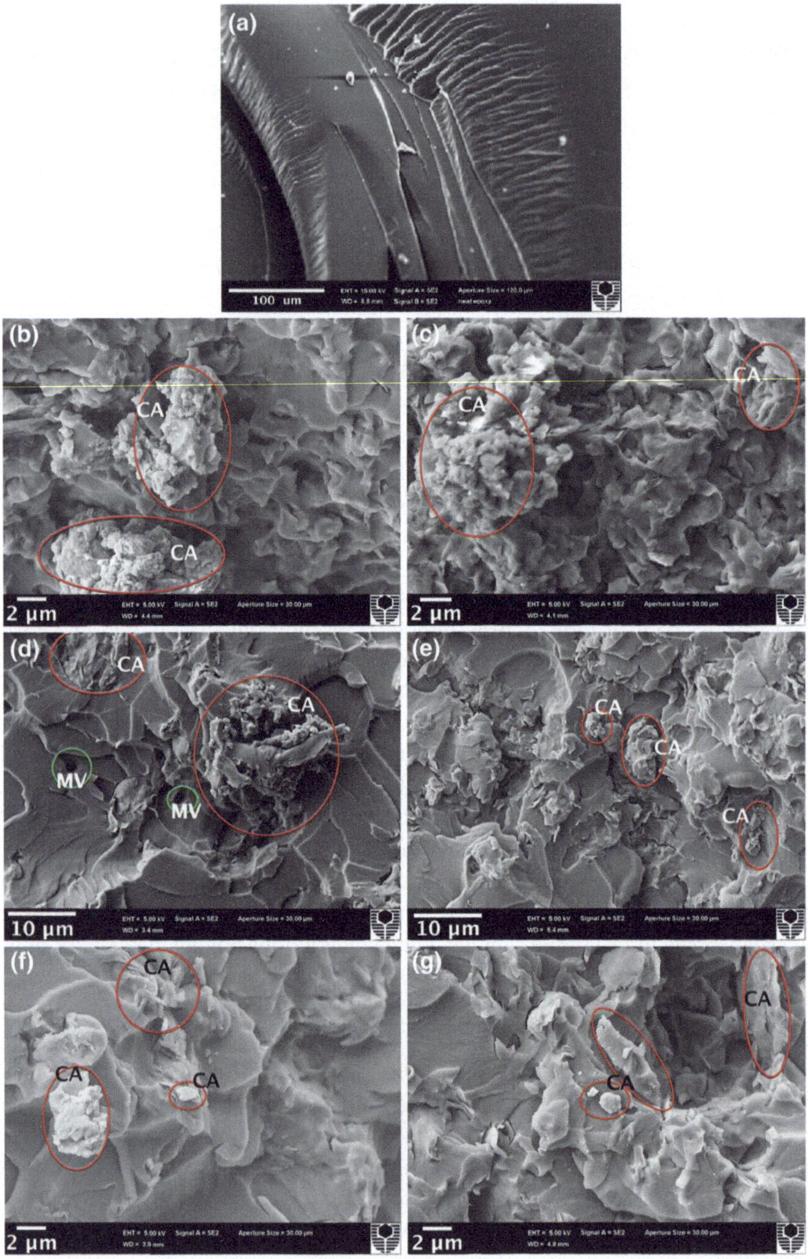

Fig. 3.3 SEM micrographs of epoxy/clay nanocomposites with different clay types and contents: **a** neat epoxy, **b** 3 wt% Cloisite 10A clays, **c** 10 wt% Cloisite 10A clays, **d** 3 wt% Cloisite 15 clays, **e** 10wt% Cloisite 15 clays, **f** 3 wt% Cloisite 93A clays, **g** 10 wt% Cloisite 93A clays, **h** 3 wt% Cloisite Na^+ clays and **i** 10 wt% Cloisite Na^+ clays. The letters CA and MV in circled areas represent clay aggregates and microvoids, respectively [1]

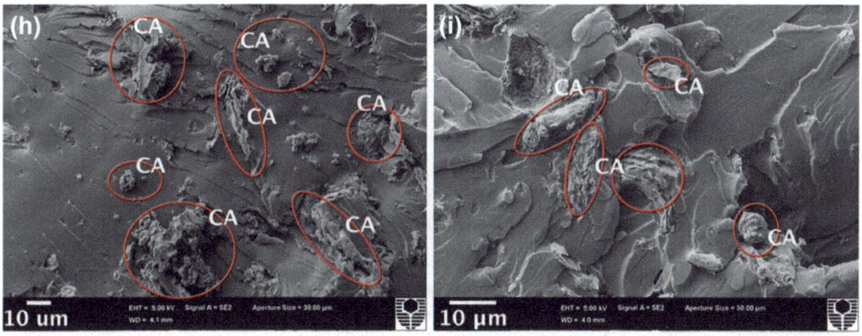

Fig. 3.3 (continued)

into microsized bubbles [11], as confirmed in Fig. 3.4. The use of inappropriate rotor speed makes it less effective to separate fillers/particles and reduce air bubbles in epoxy/clay mixtures. When the combination of three different pre-mixing processes, consisting of mechanical mixing, ultrasonication and centrifugal mixing, was directly employed in this study, it was manifested that clay wettability into polymer matrices failed to be substantially enhanced without the optimisation of material formulation and processing conditions according to resulting morphological structures in TEM and SEM analyses.

3.2 Mechanical Properties

3.2.1 Tensile Properties

Tensile properties of epoxy/clay nanocomposites with different clay types and contents prepared by three combination of premixing process were illustrated in Fig. 3.5. As demonstrated in Fig. 3.5a, it was noted that different clay types revealed lower tensile strength values as compared with that of neat epoxy at all clay content levels. The addition of 3wt% Cloisite 93A clays presented a maximum tensile strength at 43.74 MPa as opposed to other clay types and clay contents in nanocomposites, despite being still 27% lower than that of neat epoxy. In addition, with increasing Cloisite 93A clay content up to 5wt%, the tensile strength of nanocomposites declined further to the lowest value (i.e. about 51% lower than that of neat epoxy). A similar trend was also demonstrated in nanocomposites reinforced with Cloisite 15 clays. The addition of 3 wt% Cloisite 15 clays could reduce the tensile strength of nanocomposites by approximately 27% when compared with that of neat epoxy. Such a decreasing trend remained with a further increase in clay content. On the other hand, the incorporation of Cloisite 10A clays yielded low tensile strengths of approximately 30 MPa at the clay content of 3 wt%. Nonetheless, tensile strength of such nanocomposite reinforced with Cloisite 10A clays was about 60% lower than that of neat epoxy at

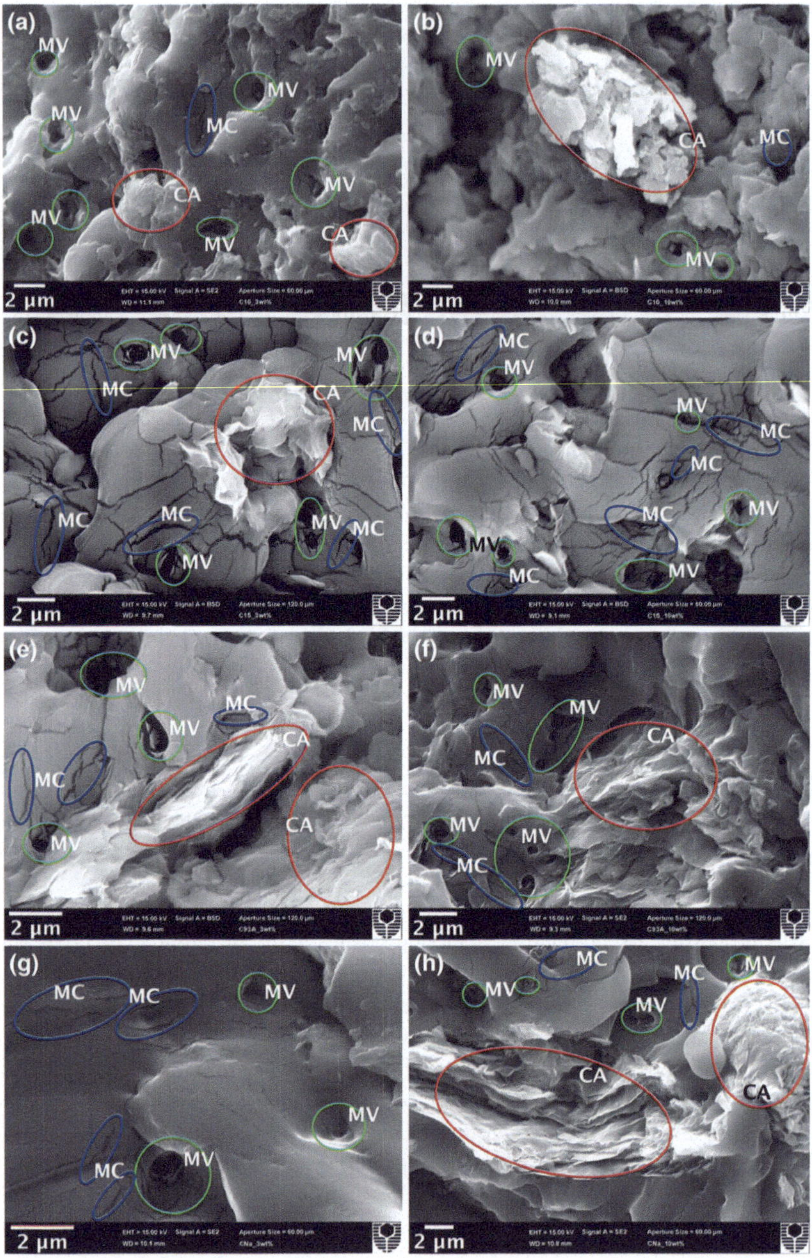

Fig. 3.4 SEM micrographs of epoxy/clay nanocomposites with different clay types and contents at high magnification levels: **a** 3 wt% Cloisite 10A clays, **b** 10 wt% Cloisite 10A clays, **c** 3 wt% Cloisite 15 clays, **d** 10 wt% Cloisite 15 clays, **e** 3 wt% Cloisite 93A clays, **f** 10 wt% Cloisite 93A clays, **g** 3 wt% Cloisite Na^+ clays and **h** 10 wt% Cloisite Na^+ clays. The letters CA, MV and MC in circled areas represent clay aggregates, microvoids and microcracks, respectively [1]

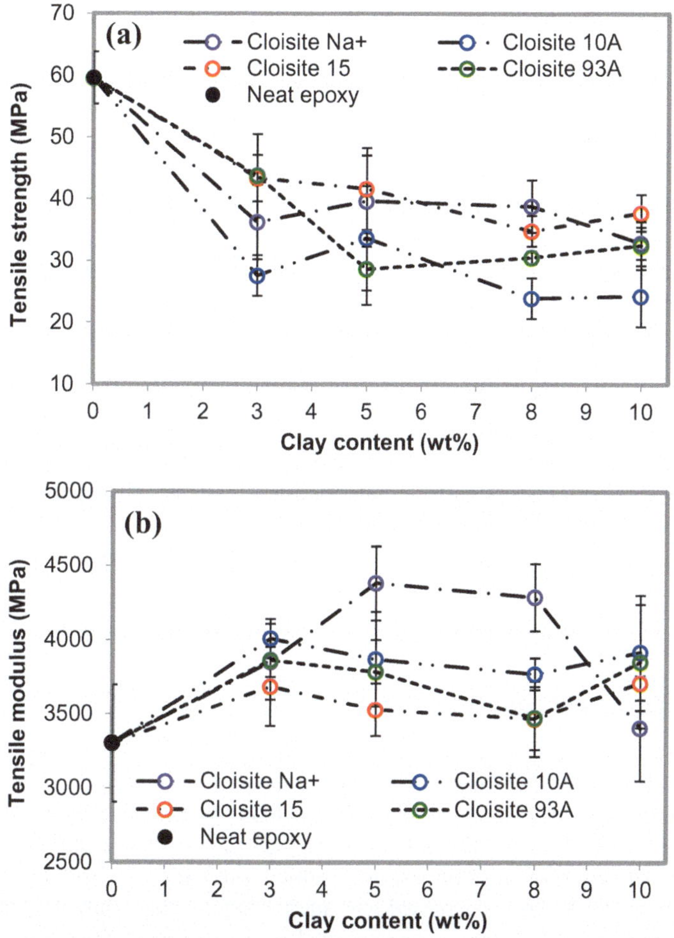

Fig. 3.5 Tensile properties of epoxy/clay nanocomposites at different clay contents: **a** tensile strength and **b** tensile modulus

all clay contents. In addition, the inclusion of pristine clay fillers (i.e. Cloisite Na^+ clays) from 3 to 10 wt% did not yield a substantial improvement in tensile strength of epoxy/clay nanocomposites. The tensile strength of such nanocomposites increased slightly by approximately 7% in a monotonic manner at Cloisite Na^+ clay contents between 3 and 8 wt%. Whereas, it was 16% decrease with the inclusion of 10 wt% clays, which was reduced to the lowest strength value. Nonetheless, tensile strengths of such nanocomposites reinforced with Cloisite Na^+ clays were found to be still lower than those of neat epoxy.

Figure 3.5b shows a clear relationship between tensile modulus and clay type and contents in epoxy/clay nanocomposites. In general, good improvements in tensile modulus of epoxy/clay nanocomposites were observed at various clay types and contents. The maximum tensile modulus of approximately 4.4 GPa was reached with the addition of 5 wt% Cloisite Na^+ clays. Meanwhile, the inclusion of 10 wt% Cloisite Na^+ clays led to the lowest increase of tensile modulus, being only 3% higher as compared with that of neat epoxy. In addition, nanocomposites reinforced with organomodified clay fillers (i.e. Cloisite 10A, Cloisite 15 and Cloisite 93A clays) nearly achieved the modulus retention with the range between 4, 3.5 and 3.7 GPa, respectively regardless of clay content.

The decrease in tensile strengths of epoxy/clay nanocomposites can be associated with the presence of microvoids and clay aggregates, as confirmed by SEM micrographs. The presence of aggregated clay and microvoids can hinder effective filler-matrix stress transfer. Besides, the poor filler dispersion within matrices at the high clay content of 10 wt%, as shown in TEM micrographs, can also worsen strength levels. Furthermore, tensile strengths of epoxy clay nanocomposites declined as compared with that of neat epoxy due to the weak interfacial bonding between fillers and matrices. Meanwhile, the enhancement of tensile modulus can be ascribed to the inclusion of clay rigid fillers with a higher modulus than that of relatively soft epoxy matrices [12]. However, the reduction in tensile modulus at the high clay content of 10 wt% can arise from the poor dispersion of clay fillers, low aspect ratios, as well as the orientation of clay fillers within matrices [13].

3.2.2 Flexural Properties

Flexural testing results for epoxy/clay nanocomposites with different clay types and contents were demonstrated in Fig. 3.6. It was noted that epoxy/clay nanocomposites possess lower flexural strength than neat epoxy. Flexural strengths of epoxy nanocomposites reinforced with organomodified clays (i.e. Cloisite 10A, Cloisite 15 and Cloisite 93A clays) decreased with increasing the clay content. Quite differently, epoxy/ Cloisite Na^+ nanocomposites had an almost constant flexural strength at approximately 62 MPa at the clay contents in range of 3–10 wt% (Fig. 3.5b). Overall, epoxy/Cloisite 15 clay nanocomposites achieved the maximum flexural strength with the inclusion of 3 wt% clays, which, however, was still 12% lower than that of neat epoxy.

On the other hand, the addition of pristine and organomodified clays into epoxy matrices could enhance flexural moduli as well. Flexural moduli of nanocomposites with the addition of 3 wt% Cloisite 15 and Cloisite Na^+ clays were significantly improved by 21 and 17%, respectively, as opposed to that of neat epoxy (Fig. 3.6a). Nevertheless, a further increase in the clay content up to 10 wt% caused the retention of flexural moduli for epoxy/ Cloisite Na^+ clay nanocomposites at approximately 3.2 GPa. Conversely, flexural moduli of epoxy/Cloisite 15 clay nanocomposites demonstrated a declining trend beyond the clay content of 3 wt%. Meanwhile, epoxy nanocomposites reinforced with Cloisite 10A and Cloisite 93A clays possessed a moderate increase in flexural modulus up to 3 wt% clay inclusions. This was followed by a slight reduction at the clay contents from 3 to 8 wt%, when compared with that

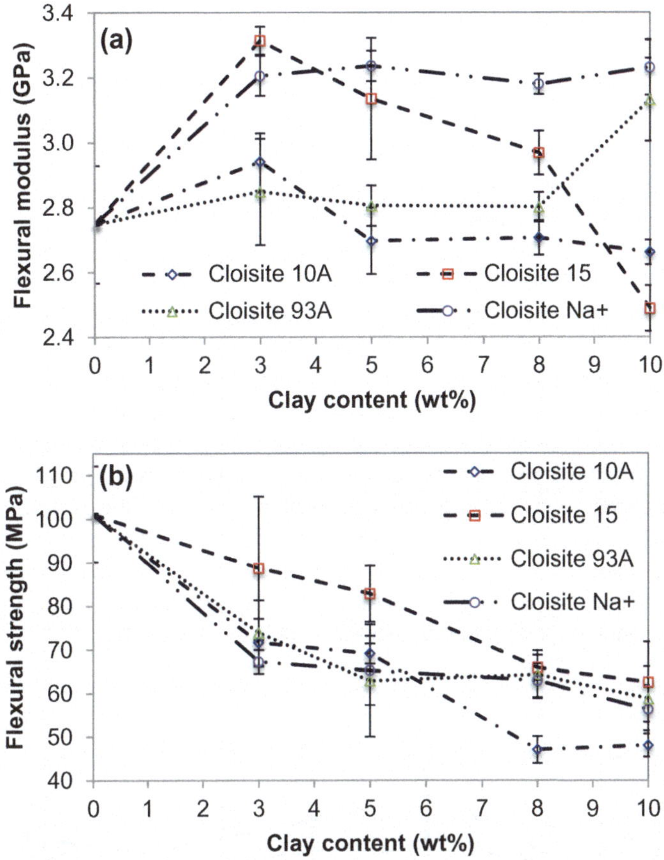

Fig. 3.6 Flexural properties of epoxy/clay nanocomposites with different clay types and contents: **a** flexural modulus and **b** flexural strength [1]

of neat epoxy. As such, it could be inferred that the addition of 3 wt% clays was an optimum level in filler content for both pristine and organomodified clays.

The reduction in flexural strength and flexural modulus for all nanocomposites was associated with many factors in material processing and morphology including clay agglomeration and poor clay wettability [14], as well as weak interfacial bonding between clay particles and epoxy matrices. In particular, a high level of clay agglomeration could hinder epoxy network junction, which adversely influenced the rigidity of nanocomposites. Low clay dispersibility and poor interfacial interaction inevitably led to the less enhancement of flexural moduli and reduced flexural strengths for epoxy/clay nanocomposites. Furthermore, the presence of microvoids detected from TEM and SEM results also indicated typical defects to initiate the crack propagation around void edges, further deteriorating flexural properties of epoxy/clay nanocomposites.

3.2.3 Charpy Impact Properties

The effect of clay type and content on impact strengths of epoxy nanocomposites was presented in Fig. 3.7. The impact strength of nanocomposites reinforced with 3 wt% Cloisite Na^+ clays was approximately 24% higher (as the highest strength level) than that of neat epoxy, as opposed to 12 and 14% increases in corresponding epoxy/Cloisite 10A clay nanocomposites and epoxy/Cloisite 93A clay nanocomposites, respectively. Afterwards, a moderate decrease in impact strengths of all nanocomposites was revealed with increasing the clay content beyond 3 wt% except for epoxy/Cloisite 10A clay nanocomposites with a significant decreasing trend up to 10 wt%. Overall, epoxy/Cloisite 15 clay nanocomposites possessed a least decreasing level of impact strengths in range from 3 to 10 wt % in clay content.

The improvement of impact strengths of nanocomposites, especially at a low clay content of 3 wt%, benefited from relatively good clay dispersion with exfoliated/intercalated structures to hinder the cracks. More well dispersed rigid clay platelets, to a certain extent, constrained internal deformation (i.e. straining effect) of epoxy matrices prone to crack initiation, resulting in much higher reinforcement efficiency. On the other hand, the presence of clay aggregates and microvoids, acting as stress concentration sites for the crack growth, eventually caused earlier impact failure of epoxy/clay nanocomposites mainly occurring at high clay contents.

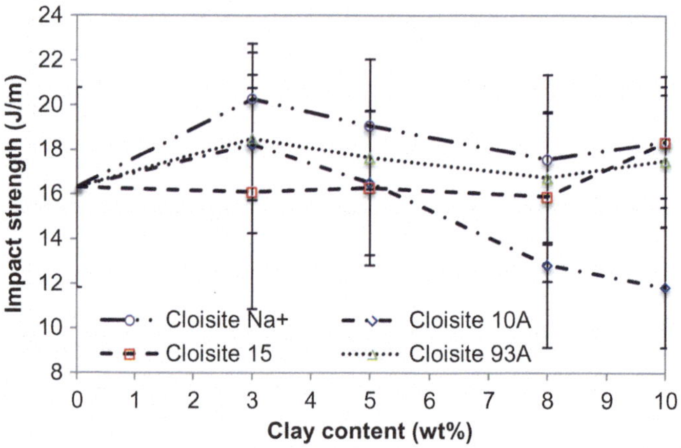

Fig. 3.7 Charpy impact strengths of epoxy/clay nanocomposites with different clay types and contents [1]

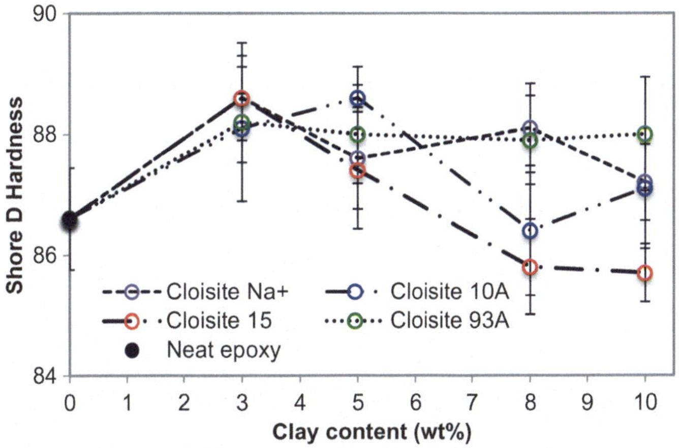

Fig. 3.8 Shore D hardness of epoxy/clay nanocomposites with different clay types and contents

3.2.4 Shore D Hardness

The effect of clay type and content on Shore D hardness of epoxy/clay nanocomposites was depicted in Fig. 3.8. In general, clay content only revealed an insignificant impact on shore D hardness, which was almost comparable to that of neat epoxy. In particular, Shore D hardness value was found to be slightly improved for about 2% at the clay contents in range from 3 to 10 wt% for nanocomposites reinforced with Cloisite Na^+, Cloisite 10A and Cloisite 93A clays when compared with that of neat epoxy. It appeared that the maximum hardness was achieved at about Shore 88.6D for epoxy nanocomposites reinforced with 3 wt% of Cloisite 15 and Cloisite Na^+ clays, as well as 5 wt% of Cloisite 10A clays. Additionally, the shore D hardness value of nanocomposite reinforced with Cloisite 15 clays showed a decreasing trend from 5 to 10 wt% in clay content. This phenomenon may be ascribed to poor clay wettability leading to agglomerated structures rather than the high levels of exfoliated/intercalated counterparts, as well as the coexistence of microvoids and pores [15].

3.3 Thermal Properties

The thermograms and glass transition temperature (T_g) values for epoxy/clay nanocomposites via differential scanning calorimetry (DSC) were depicted in Figs. 3.9 and 3.10, respectively. Overall, T_g values of epoxy/clay nanocomposites were found to be higher than that of neat epoxy, which suggested that rigid clay particles could restrict the mobility of epoxy molecules so that a higher T_g was generally required to overcome the phase change of epoxy/clay nanocomposites. The addition

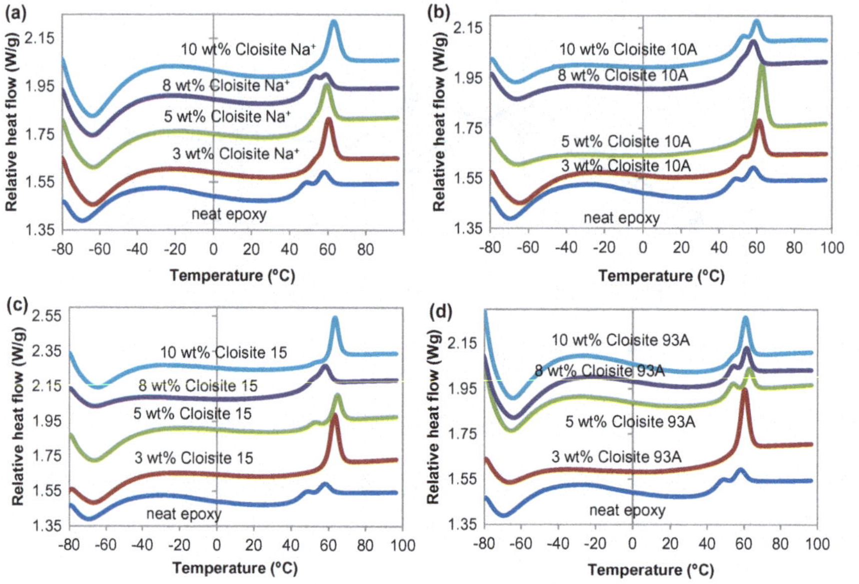

Fig. 3.9 DSC thermograms of epoxy/clay nanocomposites reinforced with **a** Cloisite Na^+ clays, **b** Cloisite 10A clays, **c** Cloisite 15 clays and **d** Cloisite 93A clays [1]

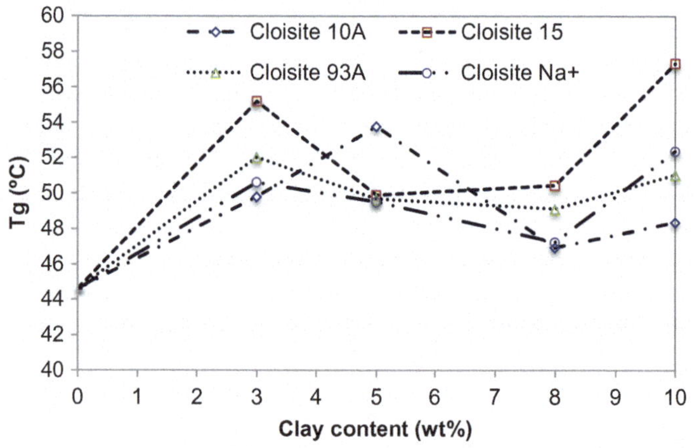

Fig. 3.10 T_g of epoxy/clay nanocomposites with different clay types and contents [1]

of 3 wt% clays in epoxy nanocomposites reinforced with Cloisite 10A, Cloisite Na^+, Cloisite 93A and Cloisite 15 clays was found to consistently increase T_g by approximately 11, 13, 16 and 23%, respectively. Even though T_g decreased slightly upon a further increase in the clay content up to 5 wt% for epoxy nanocomposites reinforced with Cloisite Na^+, Cloisite 93A and Cloisite 15 clays, associated values were still higher than that of neat epoxy. Furthermore, epoxy/Cloisite 15 clay nanocomposites revealed a dramatic increase by 28% in T_g (relative to neat epoxy) with 10 wt% clay inclusions.

These results could be interpreted by the interaction between clay particles and epoxy matrices, degree of solubility of clay particles within epoxy matrices, degree of crosslinking, chain flexibility and molecular weight [16, 17]. The decrease in T_g with higher clay contents for certain nanocomposites was ascribed to the reduction of clay dispersion levels within matrices. According to SEM results depicted in Fig. 3.4, the decrease in the chain mobility of epoxy molecules may be impeded in the case of high levels of clay aggregates and non-homogenous clay dispersion [14].

3.4 Summary

This chapter investigated the effect of material formulation of prepared epoxy/clay nanocomposites on their resulting mechanical and thermal properties using three combined pre-mixing processes, namely shear mixing, ultrasonication and centrifugation. A mix of intercalated, exfoliated and predominantly agglomerated clay structures was observed by XRD, TEM and SEM analyses. Based on the combination of three different pre-mixing processes, epoxy/clay nanocomposites possessed remarkable decreases in tensile strength (up to 60%), flexural strengths (up to 53%), which was due to weak interfacial bonding between matrices and clay fillers, as well as the presence of microvoids acting as the defects for crack initiation. Moreover, lower maximum values of flexural moduli and impact strengths were affected by poor clay wettability and aggregation according to SEM and TEM results. Finally, T_g values for nanocomposites increased regardless of clay type as compared with that of neat epoxy. This was certainly ascribed to the inclusion of rigid clays to restrict the mobility of epoxy molecules. Accordingly, it is important to investigate the preferred processing condition and material formulation on the fabrication of bioepoxy/clay nanocomposites to achieve their optimal mechanical properties, which would be mentioned in the following chapter.

References

1. Salam H, Dong Y, Davies IJ, Pramanik A (2016) The effects of material formulation and manufacturing process on mechanical and thermal properties of epoxy/clay nanocomposites. Inter J Adv ManufTechnol 87: 1999–2012.
2. Kong D, Park CE (2003) Real time exfoliation behavior of clay layers in epoxy−clay nanocomposites. Chem Mater 15 (2):419–424.
3. Wang Q, Song C, Lin W (2003) Study of the exfoliation process of epoxy–clay nanocomposites by different curing agents. J Appl Polym Sci 90(2): 511–517.
4. Akbari B, Bagheri R (2007) Deformation mechanism of epoxy/clay nanocomposite. Eur Polym J 43(3):782–788.

5. Becker O, Cheng Y-B, Varley RJ, Simon GP (2003) Layered silicate nanocomposites based on various high-functionality epoxy resins: The influence of cure temperature on morphology, mechanical properties, and free volume. Macromolecules 36(5):1616–1625.
6. Chen KH, Yang SM (2002) Synthesis of epoxy–montmorillonite nanocomposite. J Appl Polym Sci 86 (2):414–421.
7. Triantafillidis CS, LeBaron PC, Pinnavaia TJ (2002) Thermoset epoxy–clay nanocomposites: The dual role of α,ω-diamines as clay surface modifiers and polymer curing agents. J Solid State Chem 167(2):354–362.
8. Tjong SC (2006) Structural and mechanical properties of polymer nanocomposites. Mater Sci Eng R Rep 53(3–4):73–197.
9. Saber-Samandari S, Khatibi AA, Basic D (2007) An experimental study on clay/epoxy nanocomposites produced in a centrifuge. Compos Part B Eng 38(1):102–107.
10. Park JY, Davis TB, Sullivan PL (2010) Parametric study on the fabrication of clay-containing thermosetting nanocomposites. J Reinf Plast Compos 29(5): 755–770.
11. Minemura K, Murakami M (1980) A theoretical study on air bubble motion in a centrifugal pump impeller. J Fluids Eng 102(4):446–453.
12. Fu S-Y, Feng X-Q, Lauke B, Mai, Y-W (2008) Effects of particle size, particle/matrix interface adhesion and particle loading on mechanical properties of particulate–polymer composites. Compos Part B Eng 39(6): 933–961.
13. Li W, He D, Bai J (2013) The influence of nano/micro hybrid structure on the mechanical and self-sensing properties of carbon nanotube-microparticle reinforced epoxy matrix composite. Compos Part A Appl Sci Manuf 54, 28–36.
14. Bindu Sharmila TK, Ayswarya EP, Abraham BT, Sabura Begum PM, Thachil ET (2014) Fabrication of partially exfoliated and disordered intercalated cloisite epoxy nanocomposites via in situ polymerisation: Mechanical, dynamic mechanical, thermal and barrier properties. Appl Clay Sci 102: 220–230.
15. Zhang H, Popp M, Hartwig A, Madler (2012) Synthesis of polymer/inorganic nanocomposite films using highly porous inorganic scaffolds. Nanoscale 4(7): 2326–2332.
16. Alexandre M, Dubois P (2000) Polymer-layered silicate nanocomposites: Preparation, properties and uses of a new class of materials. Mater Sci Eng R Rep 28(1–2):1–63.
17. Krishnamachari P, Zhang J, Lou J, Yan J, Uitenham L (2009) Biodegradable poly(lactic acid)/clay nanocomposites by melt intercalation: A study of morphological, thermal, and mechanical properties. Int J Polym Anal Charact 14 (4):336–350.

Chapter 4
Optimisation of Material Formulation and Processing Parameters in Relation to Mechanical Properties of Bioepoxy/Clay Nanocomposites Using Taguchi Design of Experiments (DoEs)

4.1 Taguchi Method

Taguchi method is a robust design approach using an experimental design to optimise product quality with the minimal variation of quality and cost reduction. Three major steps, namely system design, parameter design and tolerance design [1] are implemented in Taguchi DoEs, amongst which the determination of functional factors or parameters belongs to the first step of system design. In the fabrication of bioepoxy/clay nanocomposites, as mentioned earlier, material formulation and processing parameters are classified into two categories of influencing factors. The second step is to select the levels of each functional factor with the minimal effect of noise data that can cause the variation of product quality and characteristics. Finally, tolerance design can employ the higher levels of material characteristics to improve product quality.

4.2 DoE Layout

In order to assess the effects of material formulation and processing parameters on tensile and impact properties of bioepoxy/clay nanocomposites, Taguchi mixed level DoEs with an L_{16} orthogonal array were used at different material selection and processing parameters, as shown in Table 4.1. Each factor comprises either two or four levels within this DoE table. The variation of factors and levels was set up according to previous research literature including ESO content [2–10], clay type [11–13], clay content [14–19], curing agent type [20–28], mechanical mixing parameters [29–33] and ultrasonication parameters [17, 34–38].

As demonstrated in Table 4.1, the explicit DoE layout consists of total 16 randomly selected experimental trials, which are very small in the number of experimental trials when compared with the full factorial DoEs including 4096 trials. The DoE responses in this study aim to maximise mechanical properties including tensile strength,

H. Salam and Y. Dong, *Bioepoxy/Clay Nanocomposites*,
https://doi.org/10.1007/978-981-16-7297-2_4

Table 4.1 L_{16} DoE layout [40]

Trial number	ESO content (wt%)	Clay type	Clay content (wt%)	Curing agent type	Mechanical mixing speed (rpm)	Mechanical mixing temperature (°C)	Mechanical mixing time (h)	Sonication frequency (kHz)	Sonication time (h)
	A	B	C	D	E	F	G	H	I
BC1	0	Cloisite Na^+	1	IPDA	200	25	1	25	1
BC2	0	Cloisite 15	5	MTHPA	500	25	1	45	3
BC3	0	Cloisite 93A	8	MTHPA	500	80	2	25	1
BC4	20	Cloisite Na^+	8	MTHPA	500	25	2	25	3
BC5	20	Cloisite 15	3	IPDA	200	25	2	45	1
BC6	60	Cloisite Na^+	5	IPDA	500	80	2	45	1
BC7	80	Cloisite 15	8	IPDA	500	80	1	25	1
BC8	0	Cloisite 10A	3	IPDA	200	80	2	45	3
BC9	80	Cloisite 93A	1	IPDA	500	25	2	45	3
BC10	80	Cloisite 10A	5	MTHPA	200	25	2	25	1
BC11	20	Cloisite 10A	1	MTHPA	500	80	1	45	1
BC12	60	Cloisite 15	1	MTHPA	200	80	2	25	3
BC13	60	Cloisite 10A	8	IPDA	500	25	1	25	3
BC14	80	Cloisite Na^+	3	MTHPA	200	80	1	45	3
BC15	20	Cloisite 93A	5	IPDA	200	80	1	25	3
BC16	60	Cloisite 93A	3	MTHPA	200	25	1	45	1

tensile modulus and impact strength of bioepoxy/clay nanocomposites. Therefore, the "larger-the-better" criterion was chosen to determine significant factors along with technical and economic considerations using the following equation:

$$S/N = -10log\left[\frac{1}{n}\sum_{i=1}^{n}\frac{1}{y_i^2}\right] \quad (4.1)$$

where *S/N* is the signal to noise ratio (dB), n is the number of measurements at each trial and y is a target response [39].

4.3 Pareto ANOVA

Pareto ANOVA was employed to identify significant factors and a preferred combination of factors in optimal conditions without the involvement of ANOVA table and F-tests. The criterion to determine significant factors is based on the overall cumulative contribution percentage over 90% [1]. In general, the cumulative addition for significant factors is determined from the highest to the lowest levels of contribution percentages. As such, the remaining factors can be taken into account as non-significant factors by concerning economical and technical aspects.

The step-by-step method to calculate contribution percentages is initiated by establishing Pareto ANOVA tables with the calculation of contribution percentages for each factor, as well as further plotting Pareto ANOVA diagrams to identify the factorial effect. In Pareto ANOVA, it should be noted that average response for each trial ($\overline{y}_i$), the response table, the total sum of squares (S_T), the sum of squares due to the mean (S_m), the sum of squares of each factor (S_i), degree of freedom of each factor (v_i), and the mean sum of squares of each factor (Mq_i) should be calculated sequentially according to the following equations to determine the contribution percentage (ρ_i) [34, 41]:

$$S_T = \sum y^2 \quad (4.2)$$

$$S_m = n\overline{y}^2 \quad (4.3)$$

$$S_i = \sum_{k=1}^{L} n_{ik}\left[\sum_{i=1}^{N}\overline{A}_{ik}^2\right] - S_m \quad (4.4)$$

$$v_i = Total\,of\,levels - 1 \quad (4.5)$$

$$Mq_i = \frac{S_i}{v_i} \quad (4.6)$$

$$\rho_i = \frac{Mq_i}{S_t} \times 100\% \tag{4.7}$$

where n_{ik} is numbers of trials at factor k and level i. n is the number of factors (i.e. $n = 9$ in this study) and L is the number of levels (i.e. mixed levels: $L = 4$ and $L = 2$ in this study). A denotes the average of *S/N* ratio of each factorial level and v_i is the degree of freedom for the given factorial level (i.e. $v_i =$ the number of level-1). S_t is the conservation of total sum of squares (i.e. $S_t = S_T - S_m$).

4.4 Confirmation Tests

To validate our DoE work, experimental confirmation tests must be conducted by using only significant factors at the optimum factorial level. Corresponding results should also be compared with the theoretical prediction of optimum conditions based on the preferred combination of factors and levels [39, 40]. The *S/N* ratio calculated in prediction tests was determined according to the following equation:

$$S/N_{Predicted} = \bar{y} + \sum_{j=1}^{k} \left(\overline{S/N}_k - \overline{S/N} \right) \tag{4.8}$$

where $\overline{S/N}_k$ is an average *S/N* ratio from determined significant factors at optimum levels, $\overline{S/N}$ is an average quantity of all *S/N* ratios, $\bar{y}$ is an average value of target response and k is the number of selected factors [40].

Furthermore, to confirm the performance of confirmation tests based on the preferred combination of DoE factors, the confidence interval (C.I) is also required to be calculated as mentioned elsewhere [39, 40], which should be generally greater than 95% in order to improve product quality from the viewpoint of engineering tests. The C.I can be determined according to the equation given below

$$C.I_{Predicted} = \sqrt{F_{\alpha,v1,v2} \times V_e \times \left[\frac{1}{n_{eff}} \right]} \tag{4.9}$$

where $F_{\alpha,v1,v2}$ is the F-ratio from the F table, α is the risk in which the confidence level is $(1 - \alpha)$ and v_1 is the degree of freedom for the numerator associated with a mean ($v_1 = 1$ in all cases). Meanwhile, v_2 is the degree of freedom for the denominator related to the degree of freedom for the pooled error variance. V_e, n_{eff} and r are the pooled error variance, the effective number of observation and the number of replicas ($r \neq 0$), respectively, in which

$$n_{eff} = \frac{total\,number\,of\,trials}{sum\,of\,degree\,of\,freedom\,use\,in\,the\,estimate\,SN\,ratio} \tag{4.10}$$

Hence, the confidence interval for a confirmation test can be given by

$$C.I_{Confirmation} = \sqrt{F_{\alpha,v1,v2} \times V_e \times \left[\frac{1}{n_{eff}} + \frac{1}{r}\right]} \tag{4.11}$$

4.5 Mechanical Properties of Bioepoxy/Clay Nanocomposites Based on Taguchi DoEs

Flexural stress–strain curves of bionanocomposites prepared by DoE method were shows in Fig. 4.1. In general, in term of ESO content, the flexural test of bionanocomposites can only be obtained up to 60 wt% ESO. Meanwhile, bionanocomposites with 80 wt% ESO could not be observed due to a very low strength value that was not detected by the load cell. This phenomenon could be affected by the addition of ESO that is likely to produce rubber-like nanocomposite materials. As previously mentioned, the plasticisation effect of additional ESO into epoxy nanocomposites system could improve the elasticity of bionanocomposites.

Additionally, for those cured by IPDA curing agent (Fig. 4.1a), it can be noted that increasing the ESO content leads to the increase in strain to failure and the decrease in flexural strength. The strains to failure for conventional nanocomposites (i.e. BC1 and BC8) and bionanocomposites (BC15) were lower than 5% of strain limit (i.e. 6.28% deflection). In addition, the flexural strain of BC5 demonstrated a yield deflection curve with the failure strain at approximately 0.18%. Meanwhile, the strains to failure of BC12 and BC16 trials were conversely over 5% of strain limit. According to ASTM D790 standard used in this testing procedure, if there is no break occurring and the strain has over the 5% strain limit, the test should be discontinued and other testing method should be taken into account.

On the other hand, similar stress–strain curves were determined in bionanocomposites cured by MTHPA (Fig. 4.1b). Increasing the ESO content leads to the improvement in strain to failure. Moreover, flexural strength and flexural modulus can also be enhanced with increasing the ESO content (i.e. BC4, BC11, BC12 and BC16) when compared with conventional nanocomposites (i.e. BC2 and BC3). Detailed flexural properties of DoE trials were listed in Table 4.2.

As illustrated in Table 4.2, flexural strengths of conventional nanocomposites cured by IPDA (i.e. BC1 and BC8) and MTHPA (i.e. BC2 and BC3) decreased as compared with those of neat epoxy. Flexural strengths of BC1 and BC8 were 26 and 21% lower, respectively as opposed to that of neat epoxy cured by IPDA, which were comparable to those cured by MTHPA (i.e. BC2 and BC3), but became 16 and 29% lower accordingly than that of neat epoxy/MTHPA. In addition, the flexural modulus

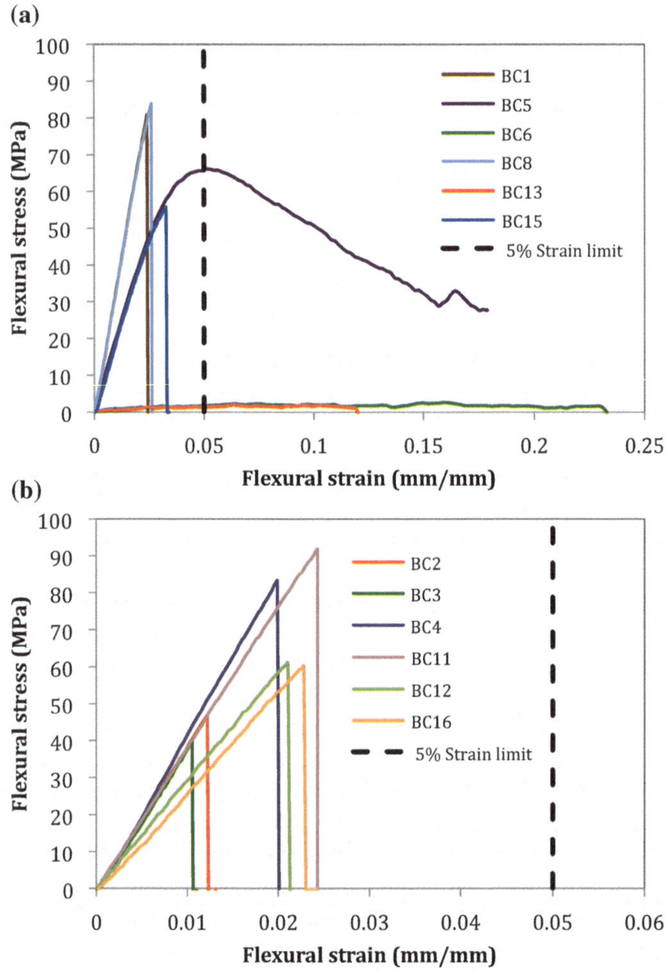

Fig. 4.1 Stress–strain curves of DoE experimental trials: **a** nanocomposites cured by IPDA, **b** nanocomposites cured by MTHPA

of conventional nanocomposites cured by IPDA was slightly lower than those of neat epoxy/IPDA (i.e. ~5% lower). However, the flexural moduli of conventional nanocomposites cured by MTHPA increased up to 14% relative to those of neat epoxy/MTHPA.

On the other hand, the functionalisation of different curing agents (i.e. IPDA and MTHPA) also affected the flexural properties of bionanocomposites. As for bionanocomposites with 20 wt% ESO, flexural strength and flexural modulus of bionanocomposites cured by IPDA were improved by approximately 300 and 600%, respectively as compared with those of bioepoxy/IPDA. However, their values were

Table 4.2 Flexural properties of DoE trials

Experimental trial	ESO (wt%)	Clay type	Clay content (wt%)	Curing agent	Flexural strength (MPa)		Flexural Modulus (GPa)	
					AVG	STDV	AVG	STDV
BC1	0	Cloisite Na^+	1	IPDA	77.79	8.19	2.04	0.23
BC8	0	Cloisite 10A	3	IPDA	82.94	6.02	2.14	0.12
NE_IPDA	0	0	0	IPDA	105.74	18.83	2.15	0.23
BC5	20	Cloisite 15	3	IPDA	60.68	5.19	1.07	0.13
BC15	20	Cloisite 93A	5	IPDA	50.27	8.51	1.20	0.21
NE_20ESO_IPDA	20	0	0	IPDA	14.04	0.67	0.16	0.05
BC6	60	Cloisite Na^+	5	IPDA	–	–	–	–
BC13	60	Cloisite 10A	8	IPDA	–	–	–	–
NE_60ESO_IPDA	60	0	0	IPDA	–	–	–	–
BC7	80	Cloisite 15	8	IPDA	–	–	–	–
BC9	80	Cloisite 93A	1	IPDA	–	–	–	–
NE_80ESO_IPDA	80	0	0	IPDA	–	–	–	–
BC2	0	Cloisite 15	5	MTHPA	45.76	4.70	2.42	0.22
BC3	0	Cloisite 93A	8	MTHPA	38.92	11.24	2.52	0.24
NE_MTHPA	0	0	0	MTHPA	54.83	16.37	2.20	0.23
BC4	20	Cloisite Na^+	8	MTHPA	68.63	21.79	3.82	0.98
BC11	20	Cloisite 10A	1	MTHPA	96.21	10.21	2.15	0.35
NE_20ESO_MTHPA	20	0	0	MTHPA	88.96	20.04	2.00	0.41
BC12	60	Cloisite 15	1	MTHPA	39.71	16.29	1.91	0.17
BC16	60	Cloisite 93A	3	MTHPA	46.95	17.31	1.57	0.40
NE_60ESO_MTHPA	60	0	0	MTHPA	–	–	–	–
BC10	80	Cloisite 10A	5	MTHPA	–	–	–	–
BC14	80	Cloisite Na^+	3	MTHPA	–	–	–	–

(continued)

Table 4.2 (continued)

Experimental trial	ESO (wt%)	Clay type	Clay content (wt%)	Curing agent	Flexural strength (MPa)		Flexural Modulus (GPa)	
					AVG	STDV	AVG	STDV
NE_80ESO_MTHPA	80	0	0	MTHPA	–	–	–	–

Note AVG = average value and STDEV = standard deviation

still lower (i.e. 27 and 44%, respectively) as opposed to those of conventional nanocomposites. In contrast, flexural strength and flexural modulus of bionanocomposites cured by MTHPA (i.e. BC11, 1 wt% C10A) revealed the improvement of approximately 8.1 and 7.7%, respectively relative to those of neat bioepoxy/MTHPA with 20 wt% ESO, as well as 37 and 50% higher accordingly when compared with those of bionanocomposites cured by IPDA (BC5). In addition, flexural strength of BC4 (i.e., bionanocomposites reinforced with 8 wt% Cloisite Na^+ clays) was found to decline being about 23% lower than that of bioepoxy/MTHPA with 20 wt% ESO even though it showed an increasing trend in flexural modulus by about 91%.

On the other hand, mechanical properties such as tensile strength, tensile modulus and impact strength of bioepoxy/clay nanocomposites prepared based on Taguchi DoEs were illustrated in Figs. 4.2, 4.3 and 4.4, respectively, along with those of neat bioepoxies. As shown in Fig. 4.2, it was noted that nanocomposites cured by IPDA possessed consistently higher tensile strengths when compared to those with MTHPA at the ESO contents of 0 and 20 wt%, respectively. A trade-off trend in relation to tensile strength was manifested when the ESO content was in range of 60–80 wt%. The highest tensile strength of nanocomposites was achieved at 49.56 MPa with the inclusion of 1 wt% Cloisite Na^+ clays and the ESO content of 0 wt% (i.e. BC1). This was followed by 41.37 MPa for nanocomposites cured by IPDA with 0 wt% ESO and 3 wt% Cloisite 10A clays (i.e. BC8). It was worth mentioning that both BC1 and BC8 belonged to conventional epoxy/clay nanocomposites without renewable biocomponents. However, their corresponding tensile strengths were still found to be 17 and 30% lower, respectively, as opposed to that of neat conventional epoxy. Nonetheless, mixed tensile results were revealed for conventional nanocomposites cured by MTHPA, as evidenced by a decrease of 28% in tensile strength for BC3 and a significant strength improvement of 69% for BC2 relative to that of corresponding epoxy. On the contrary, the lowest tensile strength was obtained for BC14 based on bioepoxy/clay cured by MTHPA with 80 wt% ESO and 3 wt% Cloisite Na^+ clays. Nonetheless, the relative strength still increased by 38% compared with that of neat bioepoxy at the same ESO content.

On the other hand, it was shown that the effect of additional ESO into epoxy/clay nanocomposites could significantly alter nanocomposite properties in comparison with that of neat bioepoxy despite the fact that such property levels were still lower than those of conventional nanocomposites. This finding could be exemplified with the strength increases of bionanocomposites by 37 and 15% at the same ESO content

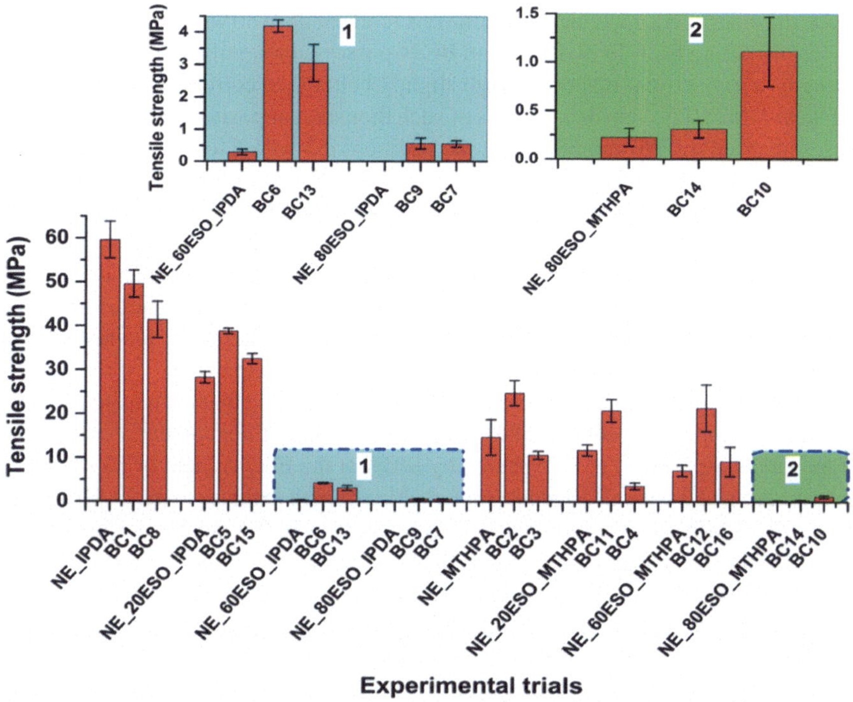

Fig. 4.2 Effects of material formulation and processing parameters on tensile strength of bioepoxy/clay nanocomposites [40]

of 20 wt% by adding 3 wt% Cloisite 15 (i.e. BC5) clays and 5 wt% Cloisite 93A (i.e. BC15) clays, respectively, as opposed to that of neat bioepoxy cured by IPDA. As for those bionanocomposites cured by MTHPA, the tensile strengths of BC4 with 8 wt% Cloisite Na^+ clays and BC11 with 1 wt% Cloisite 10A clays were determined to be 3.58 and 20.82 MPa, respectively, intending to be inferior to BC5 and BC15.

In addition, with increasing the ESO content beyond 60 wt%, tensile strengths of bionanocomposites declined by almost 99% when cured with both IPDA and MTHPA in contrast to that of corresponding conventional epoxy/clay nanocomposites. This severe detrimental effect could be attributed to ESO plasticisation effect, which gave rise to more rubbery phases in bioepoxy blends with a lower crosslinking density, as compared with conventional epoxy [4, 43, 44]. Notwithstanding a typical decreasing tendency in tensile strength for bionanocomposites with increasing the ESO content, bionanocomposites at the ESO content of 60 wt% cured by MTHPA and reinforced with 1 wt% Cloisite 15 clays (i.e. BC12) and 3 wt% Cloisite 93A clays (i.e. BC16) still possessed a significant improvement of 203 and 29% in tensile strength, respectively, when compared with that of corresponding bioepoxy. Furthermore, such strength values were found to be 410 and 198% higher than that of bioepoxy nanocomposites cured by IPDA reinforced with 8 wt% Cloisite 10A clays (i.e. BC13).

On the other hand, upon increasing the ESO content up to 80 wt%, all bionanocomposites including BC7, BC9, BC10 and BC14 possessed a significantly lower tensile strength in a consistent manner though slightly better than corresponding bioepoxy cured with MTHPA. Tensile strengths of such bionanocomposites cured by MTHPA were enhanced by 398 and 38% with the addition of 5 wt% Cloisite 10A clays (i.e. BC10) and 3 wt% Cloisite 93A clays (i.e. BC14) relative to that of corresponding bioepoxy accordingly. In addition, it was worth mentioning that bioepoxy at the ESO content of 80 wt% cured with IPDA could be successfully fabricated in material processing, but underwent the particular difficulty in conducting mechanical testing due to its highly rubber-like gel behaviour. This phenomenon would cause structural damage after demoulding in a solution casting process. Therefore, its mechanical properties were unfortunately no available for any data comparison.

In addition to the effect of ESO content, Fig. 4.2 indicated that the tensile strength of bionanocomposites could also be greatly influenced by filler content. For the case of bionanocomposites (i.e. BC5) cured by IPDA at the ESO content of 20 wt%, the incorporation of 3 wt% C15 led to a remarkable strength increase to 38.77 MPa relative to 28.23 MPa for neat bioepoxy (i.e. equivalent to 36.9% increase). Conversely, corresponding bionanocomposites (i.e. BC4) cured by MTHPA with 8 wt% Cloisite Na^+ clays underwent a strength decrease by approximately 71%. Nonetheless, at the same ESO content of 60 wt%, the highest tensile strength level was detected for MTHPA cured bionanocomposites reinforced with Cloisite 15 clays (i.e. BC12) when compared with other corresponding nanocomposites (i.e. BC6, BC13 and BC16) and neat bioepoxies. The significant improvement in tensile strength of bionanocomposites could be associated with the inclusion of well-dispersed clay particles at relatively low filler contents (i.e. 1 and 3 wt%), as well as the comparatively low ESO content used (Fig. 4.2). Apparently, increasing the ESO content tended to induce the substantial reduction in tensile strength of bionanocomposites, as mentioned earlier. As such, the increase in tensile strength was associated with proper material compatibility between bioepoxy blends and curing agents due to their reactivity to produce crosslinking bonds, as well as the effect of well dispersed rigid clay fillers in matrices [6, 20, 38].

In relation to tensile modulus of bionanocomposites, as illustrated in Fig. 4.3, overall bioepoxy/clay nanocomposites cured by MTHPA possessed much higher tensile moduli than those cured by IPDA regardless of different ESO contents used. The maximum modulus level was reached at approximately 4.46 GPa for conventional epoxy/clay nanocomposites (i.e. 0 wt % ESO) cured by MTHPA with 8 wt % Cloisite 93A clays (i.e. BC3), which was about 35% increase relative to that of corresponding neat epoxy at 3.30 GPa. Conversely, the increase in the ESO content appeared to reduce tensile modulus of bionanocomposites irrespective of different curing agents used. When the ESO content increased from 0 to 20 wt % in bionanocomposites cured by IPDA, their tensile moduli significantly decreased by the maximum level about 48% (i.e. BC5), which was followed by the worst reduction of approximately 99% at the ESO content of 80 wt % (i.e. BC7), as compared with the

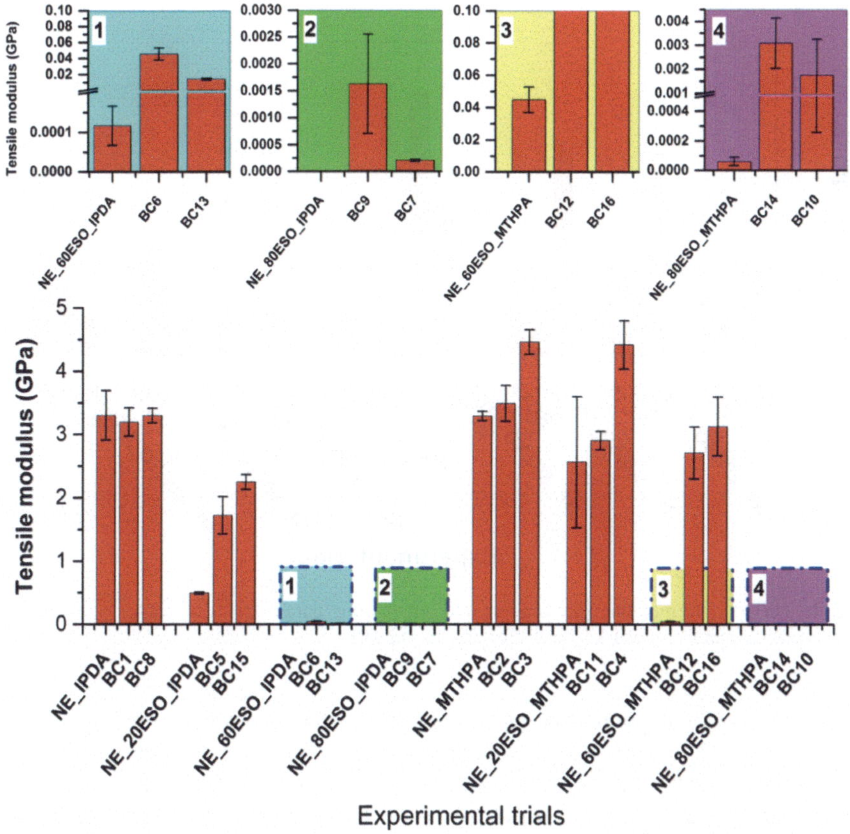

Fig. 4.3 Effects of material formulation and processing parameters on tensile modulus of bioepoxy/clay nanocomposites [40]

tensile modulus of BC8. The lowest tensile modulus was achieved for bionanocomposites (80 wt % ESO) cured by IPDA with 8 wt % Cloisite 15 (i.e. BC7) at only 0.0002 GPa among all DoE trials.

On the other hand, the addition of rigid fillers (i.e. unmodified and organomodified clays) in bionanocomposites was confirmed to enhance their tensile moduli relative to those of corresponding bioepoxies, as observed in Fig. 4.3. The addition of 5 wt % C93A in bionanocomposites cured by IPDA at the ESO content of 20 wt % (i.e. BC15) led to a modulus increase by 357% relative to that of corresponding neat bioepoxy and by 1300% with respect to BC9 at the higher ESO content of 80 wt %. As a result, significant enhancements of tensile strength and tensile moduli could be attributed to the addition of clay fillers and effective stress transfer at clay filler–matrix interfaces when rigid clay fillers were homogeneously dispersed by using preferred particle dispersion and mixing techniques [45, 46].

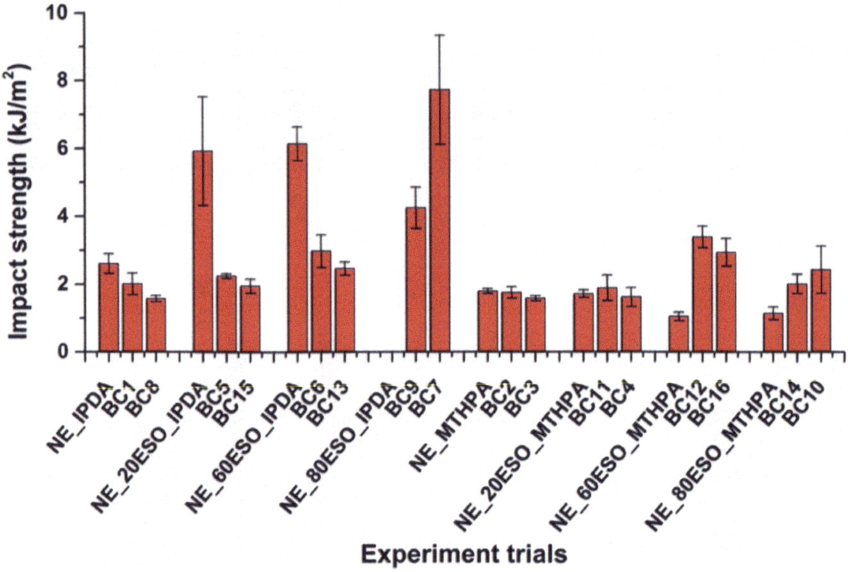

Fig. 4.4 Effects of material formulation and processing parameters on impact strength of bioepoxy/clay nanocomposites [40] (Refer Fig. 1c in [40])

With respect to the impact properties of bionanocomposites shown in Fig. 4.4, an increase in ESO content yielded an improvement in their impact strength with relatively high strength values detected especially when cured with IPDA. The maximum impact strength at 7.73 kJ/m^2 was reached for bionanocomposites cured by IPDA when reinforced with 8 wt% Cloisite 15 clays at the ESO content of 80 wt% (i.e. BC7), which was well above those of neat epoxies and bioepoxies prepared in this study.

In contrast, bionanocomposites cured by MTHPA had the greatest impact strength of 3.39 kJ/m^2 with the inclusion of 1 wt % Cloisite 15 clays at the ESO content of 60 wt % (i.e. BC12). Meanwhile, when compared with the impact strength of neat bioepoxy, a general declining trend with respect to impact strengths of bionanocomposites cured by IPDA was manifested at the ESO contents in range from 0 to 60 wt %. For instance, the impact strength of bionanocomposites in BC5 decreased by approximately 62% as opposed to that of bioepoxy with 20 wt % ESO. In contrast, a strength improvement of about 225% occurred for bionanocomposites cured by MTHPA in BC12 relative to that of bioepoxy with 60 wt % ESO, which was still much lower than those cured by IPDA (i.e. BC7 and BC9). The improvement in impact properties of bioepoxy/clay nanocomposites, as mentioned earlier, could be correlated with the increase in flexibility as a result of plasticisation effect of additional ESO [2, 22]. Nonetheless, with the inclusion of clay fillers, well-dispersed clay fillers could reduce the ductility of bioepoxy and further resulted in much stiffer bionanocomposites [5, 47]. The role of curing agent lies in its contribution to improve crosslinking density, and thus ultimately maximise load transfer and minimise crack initiation on the interfacial areas [22].

Overall, mechanical properties of conventional nanocomposites and bionanocomposites (i.e. tensile strength, tensile modulus and impact strength) depended more on the compatibility between bioepoxy matrices, clay fillers and curing agents to affect the interfacial bonding between matrices and clay fillers. However, processing parameters on manufacturing bioepoxy/clay nanocomposites also influenced the wettability of clay fillers in matrices. As such, mechanical properties of bionanocomposites could be impacted by multifaceted factors in both material formulation and processing parameters. Overall, the identification of significant factors and preferred combination of factors was essential to achieve optimal mechanical properties of bionanocomposites.

4.6 Evaluation of Significant Factors

With the aid of Pareto ANOVA, significant parameters on manufacturing bioepoxy/clay nanocomposites were identified to achieve optimum conditions for their mechanical properties based on the contribution percentage of each factor, as illustrated in Fig. 4.5a–c. Significant factors were calculated based on the cumulative contribution percentages over 90% from the highest to the lowest levels. As for tensile strengths of bionanocomposites depicted in Fig. 4.5a, factors A (ESO content), E (mechanical mixing speed) and C (clay content) were deemed to be significant with corresponding contribution percentages of 66.63, 19.09, and 7.01%. In particular, ESO content was identified as the most predominant factor among all nine factors in this DoE study. On the other hand, factors D (curing agent type), I (sonication time), H (sonication frequency), F (mechanical mixing temperature), and G (mechanical mixing time) were found to be non-significant factors with the total contribution percentage up to 7.27%. Furthermore, in response to the greatest tensile modulus level, ESO content was also determined as the most significant factor with the contribution percentage of 72.96%, which was followed by curing agent type (13%) and mechanical mixing speed (7.36%) as the second and third significant factors, respectively, Fig. 4.5b. The other six factors, namely clay content, clay type, sonication frequency, sonication time, mechanical mixing time and temperature could be categorised as non-significant factors with the total contribution percentage of only 6.68%. This result could be related to ESO plasticisation effect to influence the strength and flexibility of bionanocomposites, and higher ESO content could improve a rubber-like behaviour of bioepoxy matrices [46]. Besides, the compatibility between bioepoxy matrices and curing agents also impacted tensile properties of bionanocomposites due to their strong crosslinking for better tensile strength. Moreover, the inclusion of well-dispersed clay fillers within bioepoxy matrices enhanced both the tensile strength and rigidity of resulting bionanocomposites [46].

Conversely, as for impact strengths of bionanocomposites with Pareto ANOVA results shown in Fig. 4.5c, ESO content, curing agent type, clay type, mechanical

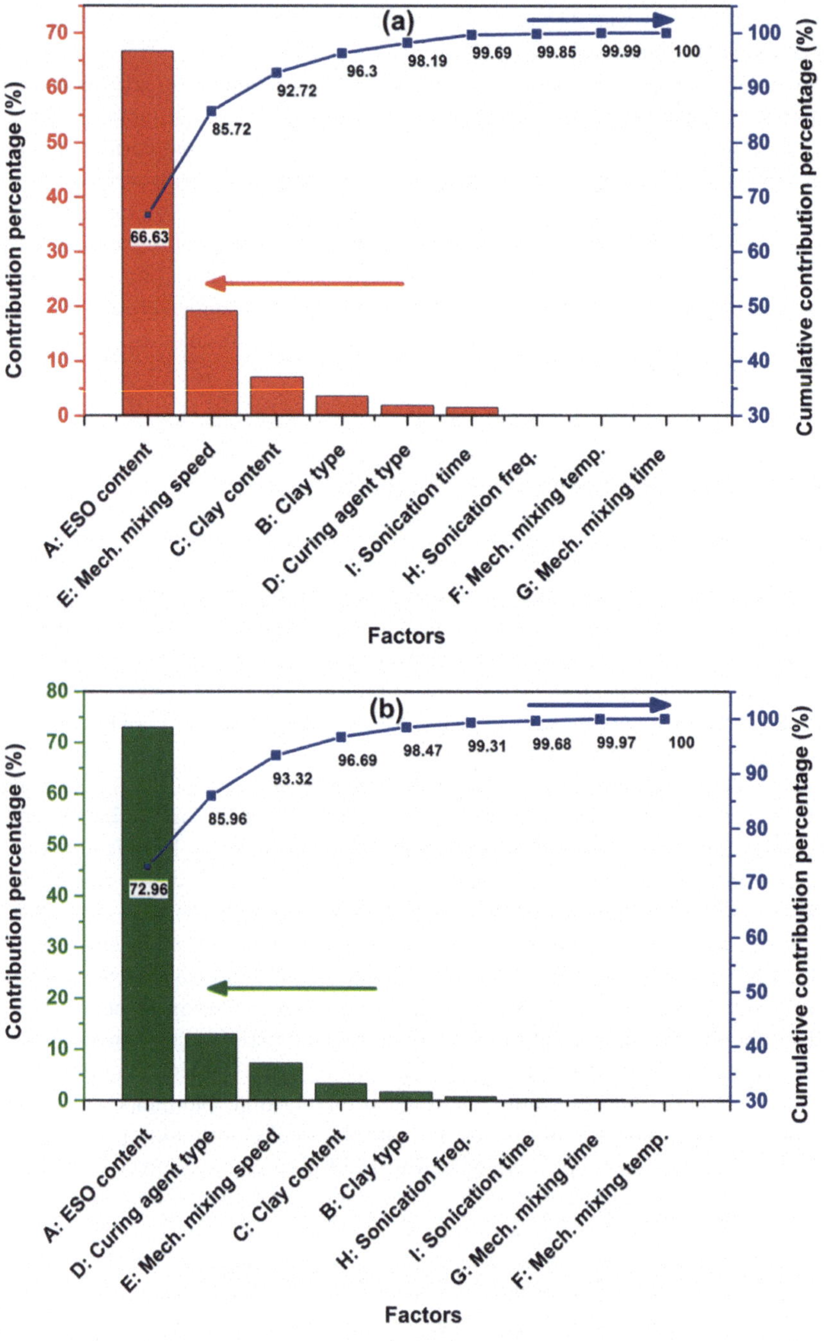

Fig. 4.5 Pareto ANOVA diagrams for the contribution percentages of mechanical properties of bioepoxy/clay nanocomposites: **a** tensile strength, **b** tensile modulus and **c** Charpy impact strength [40]

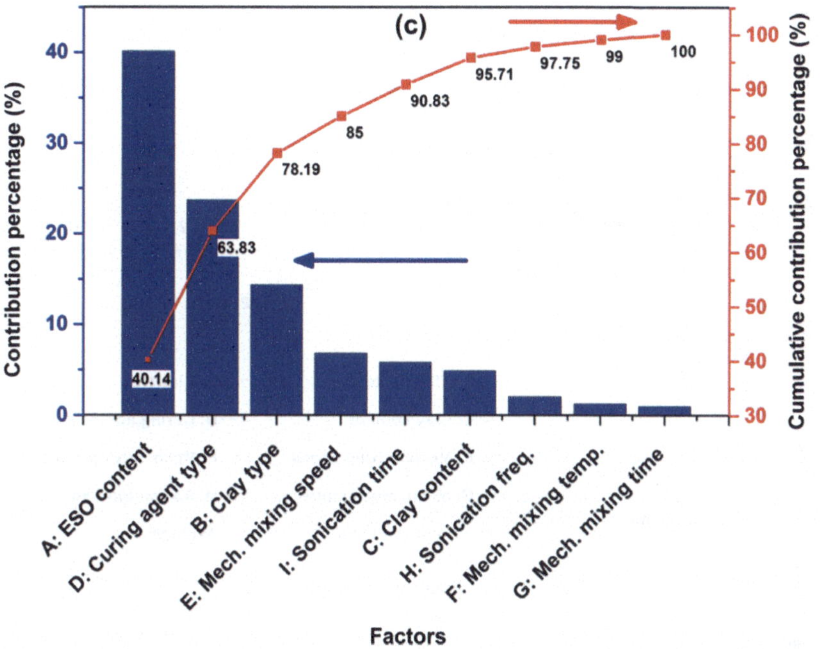

Fig. 4.5 (continued)

mixing speed and sonication time were determined to be significant factors with the contribution percentages of 40.14, 23.69, 14.37, 6.81, and 5.82%, respectively. Non-significant factors such as clay content, sonication frequency, mechanical mixing temperature and time only made a minor total contribution of 9.17%. According to the Pareto ANOVA result, it was clearly shown that the compatibility between the matrices, fillers and curing agents had significant effect on the impact strength of bionanocomposites. Additionally, well-dispersed clay fillers could improve the effective load transfer in mechanical testing. Hence, material compatibility and mixing speed were regarded as influencing factors in manufacturing bioepoxy/clay nanocomposites.

4.6.1 Preferred Combination Factors of DoEs

The sum of *S/N* ratio diagrams for each factor and level in response to different mechanical properties were illustrated in Figs. 4.6a–c. The preferred combination of factors refers to the optimal formulation based on mathematically the highest sum of *S/N* ratios for each factor. Due to typical trade-off effect commonly taking place in DoEs, it is not realistic to simultaneously achieve the maximisation of all enhanced mechanical properties based on the same optimal formulation for

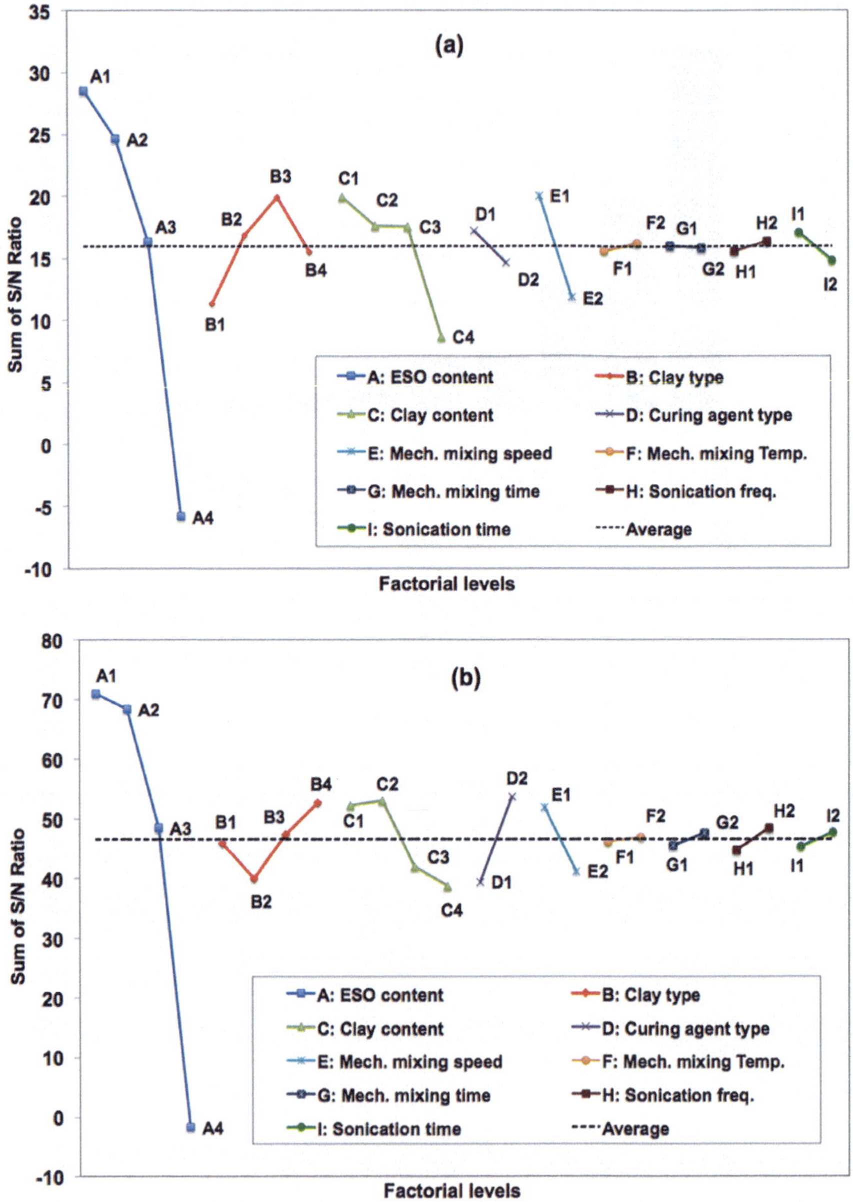

Fig. 4.6 Sum of *S/N* ratios for each factor at different levels contributing to mechanical properties of bioepoxy/clay nanocomposites: **a** tensile strength, **b** tensile modulus and **c** impact strength [40]

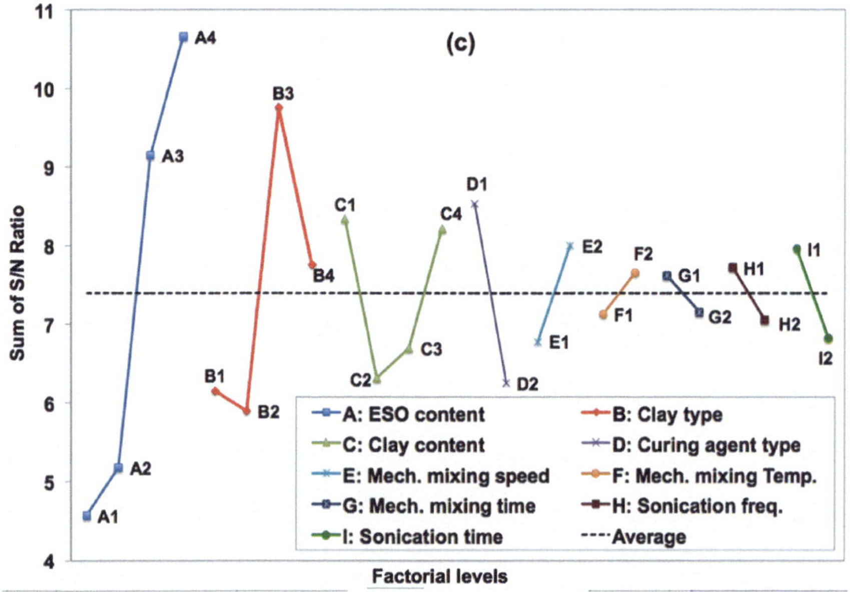

Fig. 4.6 (continued)

the purpose of achieving global optimisation [12]. Instead, a sub-optimisation process was employed in response to individual maximum mechanical properties of bionanocomposites in this study.

In general, the mathematically larger sum of *S/N* ratios led to better DoE responses. As seen in Fig. 4.6a, the preferred combination of factors was $A_1B_3C_1D_1E_1F_2G_1H_2I_1$ in order to gain the maximum tensile strength, which signified an ESO content of 0 wt%, clay type of C15, clay content of 1 wt%, curing agent type of IPDA, mechanical mixing speed at 200 rpm for 1 h, mechanical temperature at 80 °C for 1 h and sonication frequency at 45 kHz for 1 h. It was also revealed that ESO content (factor A), mixing speed (factor E), clay content (factor C) and clay type (factor B) yielded greater variation bands in terms of the sum of *S/N* ratios when compared to factors D, F, G, H and I, thus leading to a narrow range for factorial effects.

On the other hand, Fig. 4.6b presented the sum of *S/N* ratios in relation to a different range of factors in response to the highest tensile modulus, which indicated the optimal formulation of $A_1B_4C_2D_2E_1F_2G_2H_2I_2$ (i.e. ESO content of 0 wt%, clay type of C93A, clay content of 3 wt%, curing agent type of MTHPA, mechanical mixing speed at 200 rpm, mechanical mixing temperature at 80 °C for 2 h, and sonication frequency at 45 kHz for 3 h), as summarised in Table 4.3. Furthermore, the preferred combination of factors in terms of highest impact strength was also presented, namely $A_4B_3C_1D_1E_2F_2G_1H_1I_1$. Such a combination of factors for impact strength represented an ESO content of 80 wt%, clay type of C15, clay content of 1 wt%, curing agent type of IPDA, mechanical mixing speed at 500 rpm, mechanical mixing temperature at 80 °C for 1 h, and sonication frequency at 25 kHz for 1 h.

Table 4.3 Summary of preferred combination of factors based on mathematical calculation of Pareto ANOVA for DoE responses [40]

Output response	Preferred combination factors based on Pareto ANOVA calculation	Formulation composition	Final preferred combination factors	Estimate of error variance	Confirmation values (sum S/N ratio) (dB)	C.I of confirmation test at 95%	Predicted value (sum S/N ratio) (dB)	C.I of predicted value at 95%
Tensile strength	$A_1B_3C_1D_1E_1F_2G_1H_2I_1$	0 wt% ESO/C15/1 wt% C15/IPDA/500 rpm/25 °C/1 h/45 kHz/1 h	$A_1B_3C_1D_1E_1F_1G_1H_2I_1$	5.81	35.43	± 4.64	36.64	± 1.55
Tensile modulus	$A_1B_4C_2D_2E_1F_2G_2H_2I_2$	0 wt% ESO/C93A/3 wt% C15/MTHPA/200 rpm/80 °C/2 h/45 kHz/3 h	$A_1B_4C_2D_2E_1F_1G_1H_2I_1$	76.72	64.01	± 56.10	84.69	± 15.30
Impact strength	$A_4B_3C_1D_1E_2F_2G_1H_1I_1$	80 wt% ESO/C15/1 wt% C15/IPDA/500 rpm/80 °C/1 h/25 kHz/1 h	$A_4B_3C_1D_1E_2F_1G_1H_1I_1$	3.74	–	–	15.34	± 1.24

In addition, when economical and technical aspects are taken into consideration to further analyse non-significant factors, the lowest factorial levels can be employed accordingly. For example, non-significant factors such as clay type and curing agent type were selected as the optimal formulation based on the average sum of *S/N* ratios for different responses to the highest tensile strength and impact strength of bionanocomposites, including C15 for clay type and the use of IPDA as the curing agent. Other non-significant factors such as mechanical mixing temperature, mechanical mixing time, sonication frequency and time were chosen to be 25 °C for 1 h in a mechanical mixing process and 45 kHz for 1 h in subsequent sonication for all DoE responses. The summary of a preferred combination of factors based on the overall consideration of both mathematical DoE response calculations and economical and technical aspects was listed in Table 4.3, along with formulation compositions and estimates of error variance according to ANOVA.

Additionally, as mentioned previously, according to Pareto ANOVA results for the responses of tensile strength and tensile modulus in terms of ESO content, the optimum response of tensile properties could be achieved at the first level of ESO content, which was 0 wt % of ESO. Meanwhile, to fulfil the main objective to optimise material formulation and processing parameters for the manufacture of bioepoxy/clay nanocomposites, the second highest level of the sum of *S/N* ratio for ESO content (20 wt % ESO) was selected instead as the final preferred combination of factors for bioepoxy/clay nanocomposites. As mentioned earlier, the depletion of fossil fuel stock and high demand on eco-friendly materials are the main reason to functionalise such resins obtained from natural resources. By replacing the petroleum-based resin with natural counterpart, it is expected that the high demand of eco-friendly materials can be fulfilled. Accordingly, the final preferred combination of factors for bioepoxy/clay nanocomposites material with each DoE response was presented in Table 4.4.

Overall, it was worth noting that mechanical properties of bioepoxy/clay nanocomposites were more prevalently influenced by material formulation factors rather than material processing parameters. Apparently, such formulation factors consisting of ESO content, clay type and content, as well as curing agent type were

Table 4.4 Summary of confirmation tests based on final preferred combination of factors for bioepoxy/clay nanocomposites [40]

Output response	Formulation	Confirmation values (sum of S/N ratio) (dB)	C.I of confirmation test at 95%	Predicted value (sum of S/N ratio) (dB)	C.I of predicted value at 95%
Tensile strength	$A_2B_3\mathbf{C_1}D_1\mathbf{E_1}F_1G_1H_2I_1$	29.91	± 4.64	32.77	± 1.55
Tensile modulus	$A_2B_4C_2\mathbf{D_2}\mathbf{E_1}F_1G_1H_2I_1$	70.58	± 56.10	81.99	± 15.30
Impact strength	$A_2B_3\mathbf{C_1}D_1\mathbf{E_1}F_1G_1H_2I_1$	10.22	± 3.23	8.65	± 1.24

revealed to yield the total contribution percentage over 80% as opposed to 10% for certain processing parameters such as mechanical mixing speed and sonication time.

4.7 Confirmation Tests

The preferred combination of factors needed to be validated in the confirmation tests, which was then compared with statistical prediction tests. It should be noted that only significant factors were accounted for when the prediction results were determined for the optimum condition. The predicted values ($\mu_{prediction}$) of the preferred combination of factors were calculated using Eq. (4.8), along with the comparison between confirmation test data and prediction results listed in Table 4.4. It could be seen that tensile strengths of bioepoxy/clay nanocomposites for confirmation test results at 0 (CF_TS0) and 20 wt % ESO (CF_TS20) were relatively low with different levels of the sum of *S/N* ratio at approximately 1.21 and 2.86 dB, respectively. Additionally, the comparison of tensile strengths between confirmation test results (CF_TS0) and experimental trial outputs revealed that the confirmation test specimens yielded a higher tensile strength at 59.42 MPa when compared with the highest experimental trial BC1 at 49.56 MPa. As for tensile modulus, the sum of *S/N* ratios between confirmation and prediction values were approximately 20.13 and 11.41 dB lower for the confirmation tests of tensile moduli of bioepoxy/clay nanocomposites at 0 (CF_TM0) and 20 wt % ESO (CF_TM20), respectively, as opposed to those of corresponding prediction values. Moreover, a smaller value of confirmation test was also found at 0 wt % ESO (CF_TM0) as opposed to the highest tensile modulus in experimental trials (BC3), which was 2.76 GPa lower than that of BC3. This phenomenon could be affected by different clay filler contents according to Fu et al. [46]. It was noted that the addition of rigid fillers to the matrices generally could improve the modulus since the rigidity of fillers was much higher than that of organic matrices [46].

On the other hand, confirmation tests for impact strength of bionanocomposites could not be successfully conducted owing to the rubber-like gel formation of testing specimens at the ambient temperature with 80 wt% ESO for $A_4B_3C_1D_1E_2F_1G_1H_2I_1$. According to differential scanning calorimetric (DSC) observations, DSC thermograms of bionanocomposites based on the confirmation test of impact strength (CF_IS80) and BC7 exhibited two different material characteristics associated with their thermal properties, as depicted in Fig. 4.7. The glass transition temperature (T_g) of BC7 was approximately 40 °C, which was much higher than that of CF_IS80 (i.e. only about 7 °C). It was clearly stated that the T_g for BC7 was well above the ambient temperature required to ensure the full solidification of bionanocomposites. However, dominant rubbery phases of testing specimens could take place since the T_g was well below the ambient temperature. Consequently, the preferred combination of factors used in the confirmation test of impact strengths was $A_2B_3C_1D_1E_1F_1G_1H_2I_1$, which was similar to that of bionanocomposites in response to maximising tensile strength. As also shown in Table 4.4, its impact strength demonstrated a higher sum

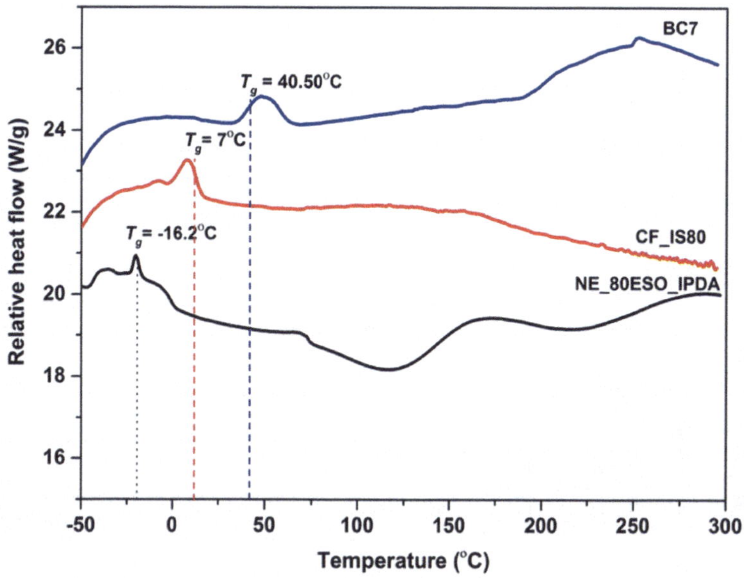

Fig. 4.7 DSC thermograms used in confirmation tests for impact strength [40]

of *S/N* ratios with a difference by 1.57 dB when compared with that of prediction tests.

Finally, the confidence interval (C.I.) was calculated to assess the quality and design robustness of the preferred combination of factors in this DoE study according to previous literature [39, 42]. As is well known within the engineering community, the 95% confidence interval should be achieved as the general principle of engineering tests to warrant the product quality and testing validity. Based on Eqs. (4.9)–(4.11) to calculate the C.I. for both confirmation and predicted results, a high degree of overlap was observed between the C.I. of confirmation tests and predicted results for all three different mechanical properties of bionanocomposites (Table 4.4), which inferred that the preferred combination of factors in terms of material formulation and processing parameters was acceptable to be used in the optimum manufacturing of bioepoxy/clay nanocomposites. It was confident that good reproducibility of such optimal bionanocomposites could be achieved through the numerical addition of significant factors predicted from ANOVA [42].

4.8 Structure–Property Relationship

Figure 4.8 demonstrated the Fourier transform infrared (FTIR) spectra of neat materials (i.e. ESO, bisphenol A diglycidyl ether (DGEBA) resin, CNa^+, and C15),

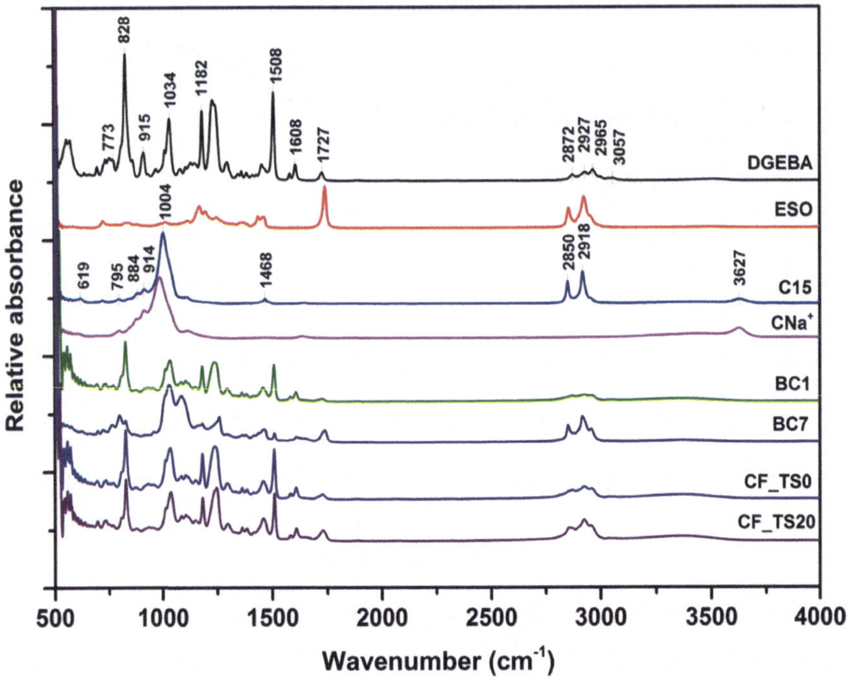

Fig. 4.8 FTIR spectra of selected neat materials and bionanocomposites. "CF_TS0" and "CF_TS20" represent confirmation tests for tensile strengths of bioepoxy/clay nanocomposites at the ESO contents of 0 and 20 wt%, respectively [40]

selected experimental trials with different ESO contents (BC1 and BC7) and confirmation test specimens in relation to tensile strength (CF_TS0 and CF_TS20). Several characteristic absorbance peaks of epoxy groups in DGEBA resin and ESO indicated the presence of rocking CH_2 at 773 cm^{-1}, stretching C–O–C of oxirane groups at 828 cm^{-1}, stretching C-O of oxirane of epoxy groups at 915 cm^{-1}, stretching C–O–C of ethers at 1034 cm^{-1}, C–C-O-C stretching at 1182 cm^{-1}, stretching C–C of aromatic at 1509 cm^{-1}, stretching $C = C$ of aromatic rings at 1608 cm^{-1}, stretching of $-C = O$ at 1727 cm^{-1}, stretching C-H of CH_2 and CH aromatic and aliphatic in range from 2872–2965 cm^{-1}, as well as stretching of C-H of the oxirane ring at 3057 cm^{-1}. Both ESO and DGEBA resin had similar peak characteristics at 1182, 1727 and 2800–2900 cm^{-1}, which were assigned to the antisymmetric stretching of ester (C–C-O-C), aliphatic $C = O$ stretching of esters and stretching C-H of CH_2 and aromatic and aliphatic CH, respectively. On the other hand, the addition of clay fillers into bioepoxy matrices exhibited the combination of absorbance peak characteristics with respect to matrices and clay fillers in bionanocomposites. In Fig. 4.8, it showed the overlap peak characteristics of bioepoxy matrices and clay fillers, which were assigned at 2918 and 2850 cm^{-1} to the asymmetry and symmetry stretching vibration of methylene groups ($-CH_2-$) and at 1004 cm^{-1} to in-plane Si–O-Si stretching, as well as at 1034 cm^{-1} to stretching C–O–C of ethers. Nonetheless, after curing agents reacted with bioepoxy/clay slurry, the FTIR spectra of experimental trial (BC1 and BC7) and confirmation test of bionanocomposites (CF_TS0 and CF_TS20) showed the reduction of epoxy peak area at 914 and 3056 cm^{-1}. The disappearance of epoxy group peaks could be associated with the opening of epoxy rings, as well as epoxy conversion after the curing process [48]. Additionally, the broadened spectra in a band range between 3200 and 3600 cm^{-1} were corresponding to OH group stretching vibration, which further confirmed the reaction of epoxy groups with curing agents to form OH groups [49]. Moreover, the presence of OH groups on the surfaces or OH groups obtained from absorbed water molecules was also clearly identified [50].

On the other hand, X-ray diffraction (XRD) and transmission electron microscopic (TEM) analyses were simultaneously used in order to correlate clay structures in bionanocomposites with resulting mechanical properties. XRD patterns of selected experimental trials, namely BC1, BC3 and BC7, as well as conventional nanocomposites and bionanocomposites based on confirmation tests of tensile strength and tensile modulus, were presented in Fig. 4.9a–c. It could be observed that the interlayer spacing (d_{001}) of Cloisite Na^{+} clays in conventional nanocomposites (BC1) shifted to a lower diffraction angle with a d_{001} increase about 0.43 nm relative to pristine Cloisite Na^{+} clay at 1.15 nm (Fig. 4.9a). Similarly, the XRD pattern of CF_TS0 reflected an increase of interlayer spacing up to 3.99 nm as opposed to 3.46 nm for Cloisite 15 clays according to Fig. 4.9b. In contrast, the (001) reflection of Cloisite 15 clays was most likely to disappear for CF_TS20 bionanocomposites with the inclusion of 1 wt% Cloisite 15 clays and 20 wt% ESO, as well as for BC7 bionanocomposites (i.e. 8 wt% Cloisite 15 clays and 80 wt% ESO). This disappearance suggested that either complete exfoliation or highly random orientation of clay fillers may occur within bioepoxy matrices. On the other hand, the XRD pattern of Cloisite 93A clays in those of confirmation tests for tensile modulus (i.e. CF_TM0 and CF_TM20) with the inclusion of 3 wt% Cloisite 93A clays did not exhibit any diffraction pattern within the diffraction-angle range, as depicted in Fig. 4.9c. On the contrary, the XRD pattern of BC3 (8 wt% Cloisite 93A clays) yieleded a significant increase in interlayer spacing by approximately 52% relative to that of Cloisite 93A clays (d_{001} = 3.92 nm), suggesting the existence of intercalated clay structures. Meanwhile, the disappearance of the (001) reflection might be related to peak broadening or low diffraction intensity arising from disordered clay orientation. This disordered orientation might be attributed to an inhomogeneous mix of intercalated and exfoliated clay structures, as well as clay aggregates. Hence, it is interesting to note that the functionalisation of nanocomposites at the ESO content of 20 wt% may improve clay dispersion within the matrices.

With respect to the wettability status, TEM analysis should be used as a supplementary characterisation technique to investigate the morphological structures and confirm the clay dispersion status in bioepoxy matrices especially when the mechanical mixing speed of 200 rpm was chosen. Figure 4.10a–d depicted TEM images of bionanocomposites in responses to final preferred combination of factors in relation to tensile strength and impact strength with similar material compositions (i.e. 20wt% of ESO with 1wt% Cloisite 15 clays and the use of IPDA), as well as tensile

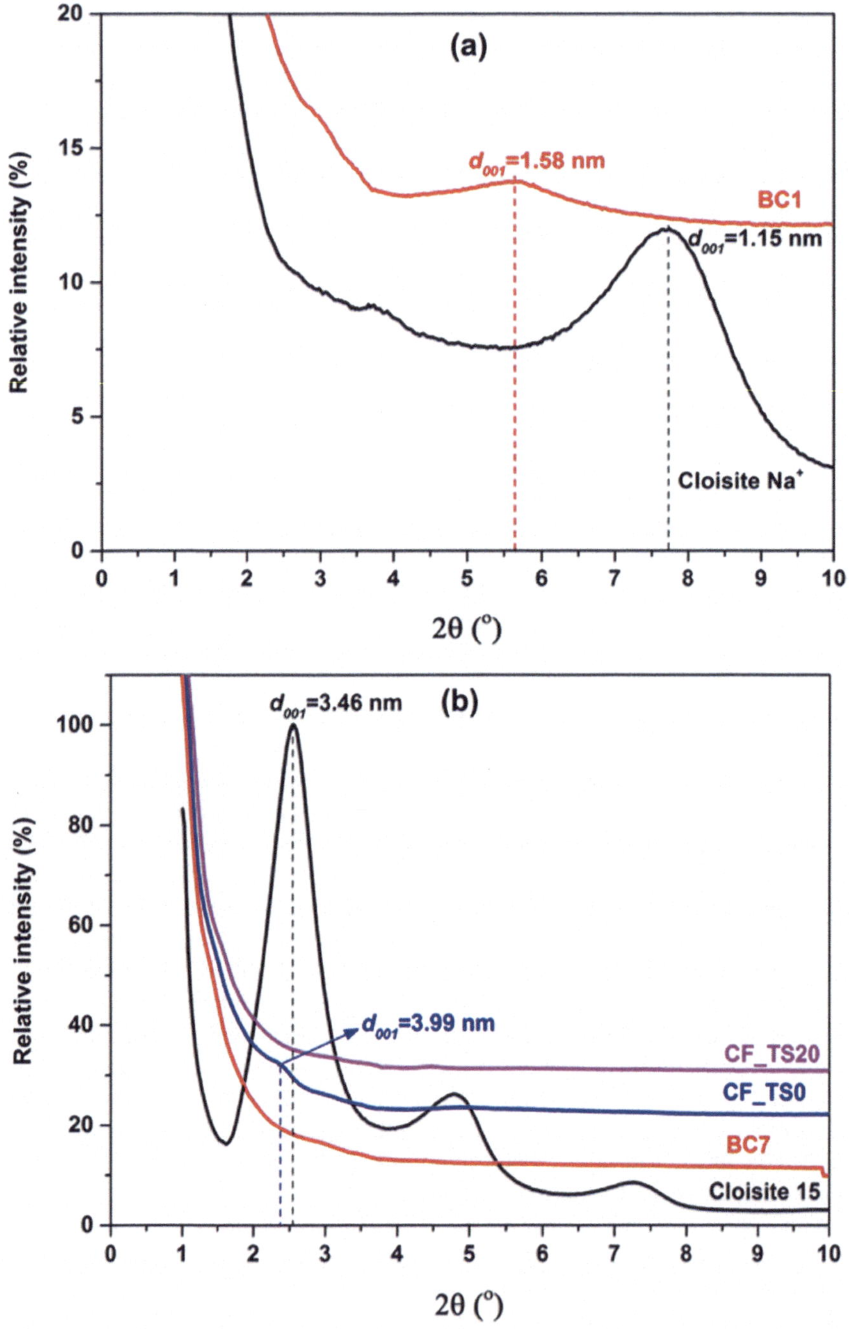

Fig. 4.9 XRD patterns of DoE trials and confirmation tests of bioepoxy/clay nanocomposites reinforced with: **a** Cloisite Na^{+} clays, **b** Cloisite 15 clays and **c** Cloisite 93A clays [40]

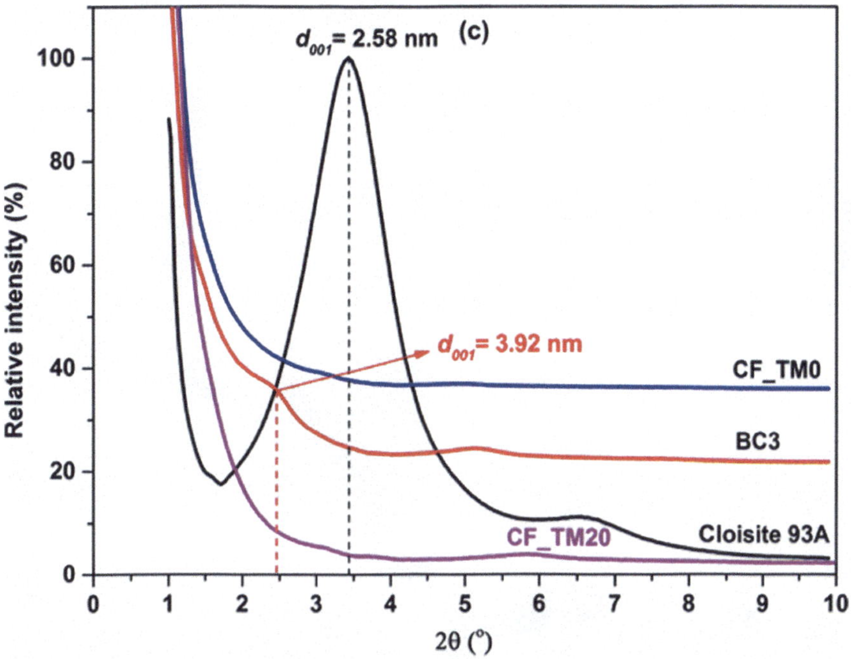

Fig. 4.9 (continued)

modulus with the material composition of 20 wt% ESO and 3 wt% Cloisite 93A clays and the use of MTHPA at different magnifications. The dark lines and clouds represented clay platelet layers and clay aggregates, respectively while the bright areas signified the existence of polymer matrices. As indicated in Fig. 4.10a, the addition of 1 wt% clays on the final preferred combination of factors in response to tensile strength (CF_TS20) and impact strength (CF_IS20) gave rise to a low level of clay wettability within such fabricated bionanocomposites. The TEM image of CF_TS20 demonstrated localised clay aggregates at about 2–3 μm with the dominant presence of small clay aggregates below 1 μm, as illustrated in Fig. 4.10a.

On the other hand, TEM images of CF_TM20 in response to tensile modulus, as depicted in Fig. 4.10c–d, respectively, suggested clay platelet layers were moderately dispersed within epoxy matrices with a mix of aggregated and intercalated clay structures. The TEM images from the confirmation test (CF_TM20) showed dominantly intercalated clay structures rather than small clay aggregates, resulting in better clay wettability when compared with CF_TS20 (Fig. 4.9a). In addition, TEM images at higher magnifications suggested the presence of exfoliation of individual clay platelets in some regions close to intercalated and agglomerated clay structures for CF_TS20 (Fig. 4.10b) and CF_TM20 (Fig. 4.10d). Hence, the use of mechanical mixing speed at 200 rpm for manufacturing bionanocomposites (CF_TS20 and CF_TM20) yielded their morphological structures that can be justified as partially exfoliated, disorderly intercalated and aggregated clay structures.

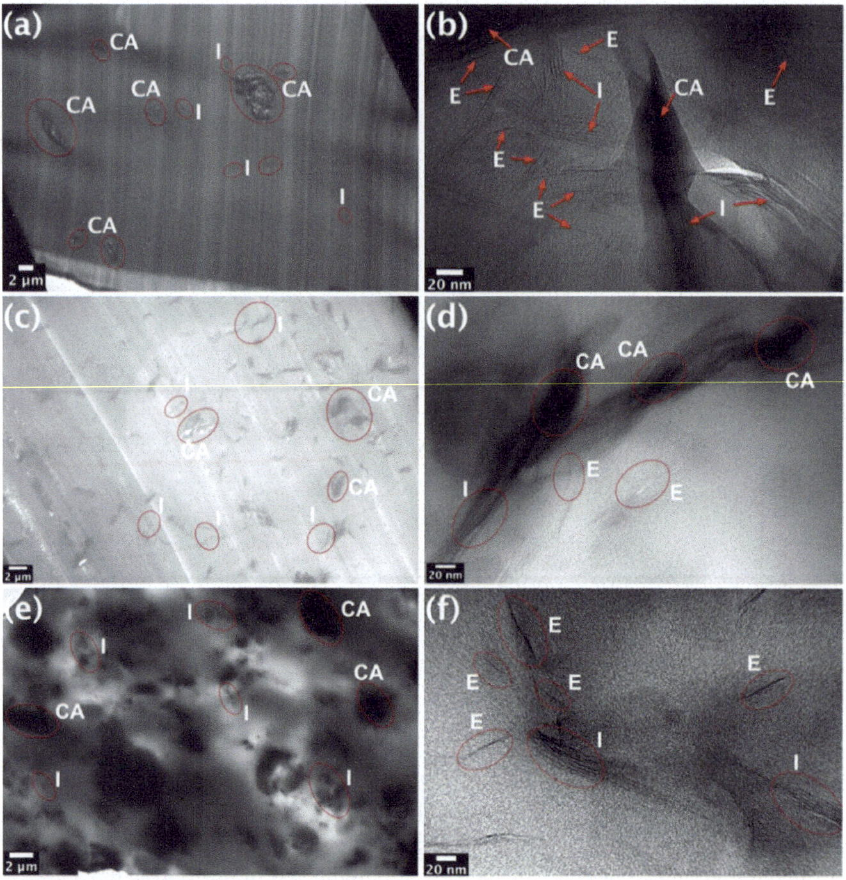

Fig. 4.10 Clay dispersion status of final preferred combination factors of bioepoxy/clay nanocomposites at different magnifications using TEM analysis: **a** tensile strength sample (CF_TS20) at 500 × and **b** at 30,000 × ; **c** tensile modulus sample (CF_TM20) at 500 × and **d** at 30,000 × , as well as **e** Charpy impact strength sample (BC7) at 500 × and **f** at 30,000 × . The letters CA, I and E represent clay aggregates, intercalated and exfoliated structures, respectively [40]

Mechanical properties of bionanocomposites also exhibited a good correlation with the fracture surface morphology. The SEM micrographs of bionanocomposites from the confirmation tests in response to their tensile strength and tensile modulus could be seen in Fig. 4.11a–d, respectively. It was observed that fracture surfaces of bionanocomposites became increasingly rougher due to clay inclusions. The fracture surface of CF_TS20, as shown in Fig. 4.11a, exhibited less rough surface characteristics relative to CF_TM20 according to Fig. 4.11c. This phenomenon was ascribed to different clay types being used in the nanocomposite systems (Cloisite 15 clays for CF_TS20 and Cloisite 93A clays for CF_TM20) and at various clay contents (1 wt% for CF_TS20 and 3 wt% for CF_TM20). Additionally, on fracture surfaces

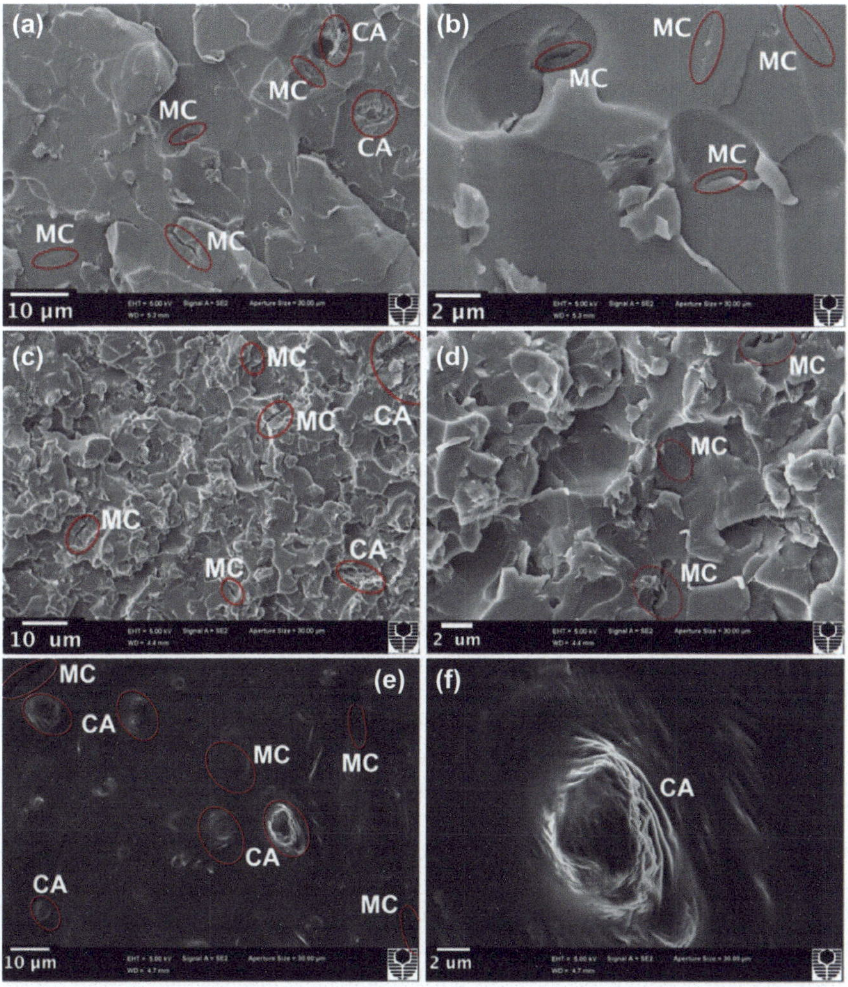

Fig. 4.11 SEM micrographs of final preferred combination factors of bioepoxy/clay nanocomposite specimens at different magnifications: **a** tensile modulus sample (CF_TS20) at 1,000 × and **b** at 10,000 × ; **c** tensile modulus sample (CF_TM20) at 1,000 × and **d** at 10,000 × , as well as **e** Charpy impact strength sample (BC7) at 1,000 × and **f** at 10,000 × . The letters CA and MC represent clay aggregates and microcracks, respectively [40]

of both specimens, it was clearly shown that clay fillers were well dispersed into the matrices with relatively small sizes of clay particles formed along the fracture surfaces. The decrease in particle size may promote the improvement of interfacial bonding between clay fillers and matrices due to resulting higher aspect ratios and specific surface areas of nanofillers to improve the mechanical strengths of bionanocomposites [45]. On the other hand, microcracks were evidently observed at the interfacial areas of clay platelets, as well as the outer faces of clay aggregates near epoxy matrices. The incorporation of organomodified clay fillers in the confirmation tests was believed to improve clay wettability and interfacial bonding between clay fillers and epoxy matrices in contrast with those based on pristine clays. This was most likely due to the presence of alkylammonium ions in clay interlayers for the improvement of clay wettability [13, 51].

With respect to impact strength, the result for the highest impact strength is in an opposite trend to that for the tensile strength counterpart. Bionanocomposites with a high impact strength are tougher materials with an enhanced ability to absorb impact energy. The addition of ESO, which is usually used as a plasticiser to modify material properties in polymer blends, could improve the toughness of bionanocomposites by absorbing more energy during impact tests [52]. This observation also suggested that the addition of ESO could enhance the interlayer spacing of clay platelets in line with the improvement of clay wettability, as confirmed by an XRD pattern and TEM images of BC7 in Figs. 4.9b and 4.10e and f, respectively. Moreover, TEM images revealed good clay wettability in Fig. 4.9e and showed the combination of randomly exfoliated and intercalated structures with clay aggregates in small sizes (Fig. 4.10f). The characteristic fracture surfaces of BC7 with the highest impact strength based on DoE trials were presented in Fig. 4.11e and f. It was clearly seen that the failure occurred for the surface morphology of bioepoxy/clay nanocomposites after tensile tests. Figure 4.11e showed finer fracture surface finish for bionanocomposites, as evidenced by good blending of ESO molecules with epoxy resin. It was also observed that the distribution of clay aggregates in the range of 2–10 μm was identified on fracture surfaces. Moreover, at a relatively high magnification (Fig. 4.11f), the SEM image illustrated how the molecules of bioepoxy matrices diffused inside the clay interlayers with clear interfacial bonding between clay fillers and matrices.

Therefore, as far as Taguchi DoE was concerned, the ESO content was determined to be the most significant factor for maximising mechanical properties of bionanocomposites such as tensile strength, tensile modulus and Charpy impact strength. An increase in ESO content yielded a significant reduction in tensile strength and tensile modulus, along with an opposite trend for impact strength. The decline of tensile strength and modulus could be attributed to ESO plasticisation effect, which might produce additional elastically rubbery phases within bioepoxy matrices with a lower crosslinking density as opposed to that of conventional epoxy [4, 44, 47].

4.9 Summary

In this study, the effects of material formulation and processing parameters were successfully investigated based on Taguchi DoEs with a mixed factorial level. The preferred combination of factors for manufacturing bioepoxy/clay nanocomposites was determined to achieve optimal mechanical properties in relation to tensile strength, tensile modulus and impact strength by means of Pareto ANOVA. SEM, XRD and TEM analyses were employed to investigate fracture surfaces of micro/nanostructures, clay wettability and dispersibility in morphological structures of resulting bionanocomposites. It was consistently indicated that ESO content was the most significant factor to maximise overall mechanical properties of bionanocomposites. The increase in ESO content was found to enhance impact strengths of bionanocomposites, especially for those cured by IPDA. Moreover, the addition of ESO also evidently yielded significant reductions in their tensile strength and tensile modulus due to ESO plasticisation effect, thus decreasing the rigidity of bioepoxy/clay nanocomposites.

Among the processing parameters, mechanical mixing speed was detected as the most significant factor though its increase did not allow for any significant improvement in tensile properties of bionanocomposites. Clay content was shown to play an important role with respect to tensile strength, being the third significant factor following ESO content and mixing speed. However, this was not the case for their tensile modulus (influenced by curing agent type as the second significant factor) and impact strength (affected by curing agent type, clay type, mixing speed and sonication time sequentially). The remaining factors comprising mechanical mixing temperature and time, as well as sonication frequency, could be categorised as non-significant factors owing to their cumulative contribution percentage of less than 10%. Notwithstanding that test confirmation results based on a preferred combination of factors in relation to tensile strength and tensile modulus were slightly lower than statistically predicted values, the confidence interval obtained at 95% was satisfactory for the validity of this DoE study. On the other hand, due to the presence of very soft phases of bioepoxy/clay nanocomposites at the ESO content of 80 wt% at ambient temperature, the confirmation test of impact strength could not be achieved. FTIR results revealed that epoxy matrices reacted with clay fillers that were assigned to stretching Si–O-Si in a range from 1000 to 1040 cm^{-1}. This indicated that silicone in clay fillers could react with oxygen and carbon in epoxy matrices. Material characterisation regarding XRD patterns, TEM and SEM observations revealed a mix of partially exfoliated, intercalated and aggregated clay structures for such bionanocomposites in both DoE trials and confirmation tests. This finding confirmed the complexity of manufacturing bionanocomposites to achieve effective clay dispersibility and simultaneously maximise their mechanical properties.

References

1. Park SH (1996) Robust design and analysis for quality engineering. Chapman and Hall: London.
2. Ratna D, Banthia AK (2000) Epoxidised soybean oil toughened epoxy adhesive. J Adhes Sci Technol 14(1): 15–25.
3. Zhu J, Chandrashekhara K, Flanigan V, Kapila S (2004) Curing and mechanical characterisation of a soy-based epoxy resin system. J Appl Polym Sci 91(6): 3513–3518.
4. Altuna FI, Espósito LH, Ruseckaite RA, Stefani PM (2011) Thermal and mechanical properties of anhydride-cured epoxy resins with different contents of biobased epoxidised soybean oil. J Appl Polym Sci 120(2): 789–798.
5. Miyagawa H, Misra M, Drzal LT, Mohanty AK (2005) Novel biobased nanocomposites from functionalised vegetable oil and organically-modified layered silicate clay. Polymer 46(2): 445–453.

6. Shabeer A, Sundararaman S, Chandrashekhara K, Dharani LR (2007) Physicochemical properties and fracture behavior of soy-based resin. J Appl Polym Sci 105(2): 656–663.
7. Jin F-L, Park S-J (2008) Thermomechanical behavior of epoxy resins modified with epoxidised vegetable oils. Polym Int 57: 577–583.
8. Raghavachar R, Sarnecki G, Baghdachi J, Massingill J (2000) Cationic, thermally cured coatings using epoxidised soybean oil. J Coat Technol 72(909), 125–133.
9. Mustata F, Tudorachi N, Rosu D (2011) Curing and thermal behavior of resin matrix for composites based on epoxidised soybean oil/diglycidyl ether of bisphenol A. Compos Part B: Eng 42(7):1803–1812.
10. Roudsari GM, Mohanty AK, Misra M (2014) Study of the curing kinetics of epoxy resins with biobased hardener and epoxidised soybean oil. ACS Sustain Chem Eng 2(9): 2111–2116.
11. Sermsantiwanit K, Phattanarudee S (2012) Preparation of bio-based nanocomposite emulsions: Effect of clay type. Prog Org Coat 74(4): 660–666.
12. Dong Y, Bhattacharyya D (2008) Effects of clay type, clay/compatibiliser content and matrix viscosity on the mechanical properties of polypropylene/organoclay nanocomposites. Compos Part A: Appl Sci Manuf 39(7): 1177–1191.
13. Kornmann X, Lindberg H, Berglund LA (2001) Synthesis of epoxy–clay nanocomposites: influence of the nature of the clay on structure. Polymer 42(4):1303–1310.
14. Liu Z, Erhan SZ (2008) "Green" composites and nanocomposites from soybean oil. Mater Sci Eng A 483–484: 708–711.
15. Tanrattanakul V, Saithai P (2009) Mechanical properties of bioplastics and bioplastic–organoclay nanocomposites prepared from epoxidised soybean oil with different epoxide contents. J Appl Polym Sci 114(5): 3057–3067.
16. Bhuyan S, Sundararajan S, Lu Y, Larock RC (2010) A study of the physical and terminological properties of biobased polymer–clay nanocomposites at different clay concentrations. Wear 268(5–6): 797–802.
17. Wang R, Schuman T, Vuppalapati RR, Chandrashekhara K (2014). Fabrication of bio-based epoxy-clay nanocomposites. Green Chem 16(4): 1871–1882.
18. Albayrak Ö, Şen S, Çaylı G, Ortaç B (2013) Bio-based polymer nanocomposites based on layered silicates having a reactive and renewable intercalant. J Appl Polym Sci 130(3): 2031–2041.
19. Lu J, Hong CK, Wool RP (2004) Bio-based nanocomposites from functionalized plant oils and layered silicate. J Polym Sci Part B: Polym Phys 42(8):1441–1450.
20. Gupta AP, Ahmad S, Dev A (2010). Development of novel bio-based soybean oil epoxy resins as a function of hardener stoichiometry. Polym Plast Technol Eng 49(7): 657–661.
21. Miyagawa H, Mohanty A, Drzal LT, Misra M (2004) Effect of clay and alumina-nanowhisker reinforcements on the mechanical properties of nanocomposites from biobased apoxy: A comparative study. Ind Eng Chem Res 43(22): 7001–7009.
22. Miyagawa H, Mohanty AK, Misra M, Drzal LT (2004) Thermo-physical and impact properties of epoxy containing epoxidised linseed oil, 1.Macromol Mater Eng 289(7): 629–635.
23. Chen Y, Xi Z, Zhao L (2016) New bio-based polymeric thermosets synthesized by ring-opening polymerisation of epoxidised soybean oil with a green curing agent. Eur Polym Int 84: 435–447.
24. Frischinger I, Dirlikov S (1993). Toughening of epoxy resins by epoxidised soybean oil. In: Keith R, Anthony JK (Eds), Toughened Plastics I, Vol. 233, American Chemical Society: Washington, 451–489.
25. Karger-Kocsis J, Grishchuk S, Sorochynska L, Rong MZ (2014) Curing, gelling, thermomechanical, and thermal decomposition behaviors of anhydride-cured epoxy (DGEBA)/epoxidised soybean oil compositions. Polym Eng Sci 54(4): 747–755.
26. Samper MD, Fombuena V, Boronat T, García-Sanoguera D, Balart R (2012) Thermal and mechanical characterisation of epoxy resins (ELO and ESO) cured with anhydrides.J Am Oil Chem Soc, 89(8): 1521–1528.
27. Liu ZS, Erhan SZ, Calvert PD (2004) Solid freeform fabrication of epoxidised soybean oil/epoxy composites with di-, tri-, and polyethylene amine curing agents. J Appl Polym Sci 93(1): 356–363.

28. Espinoza-Perez JD, Nerenz BA, Haagenson DM, Chen Z, Ulven CA, Wiesenborn DP (2011) Comparison of curing agents for epoxidised vegetable oils applied to composites. Polym Compo 32(11) 1806–1816.
29. Shi Y, Kanny K, Jawahar P (2009) Hybrid nanocomposites: Processing and properties. Adv Compos Mater 18(4): 365–379.
30. Alateyah AI, Dhakal HN, Zhang ZY (2013) Processing, properties and applications of polymer nanocomposites based on layer silicates: A review. Adv Polym Technol 32(4): 21368.
31. Atiemo-Obeng VA, Penney W, Armenante PM (2004) Solid-liquid mixing. In: Paul EL, Atiemo-Obeng VA, Kresta SM (Eds), Handbook of industrial mixing: Science and practice. John Wiley and Sons: Hoboken, 543–584.
32. Park JY, Davis TB, Sullivan PL (2010) Parametric study on the fabrication of clay-containing thermosetting nanocomposites. J Reinf Plast Compos 29(5): 755–770.
33. Gupta N, Lin TC, Shapiro M (2007) Clay-epoxy nanocomposites: Processing and properties. JOM 59(3): 61–65.
34. Bensadoun F, Kchit N, Billotte C, Bickerton S, Trochu F, Ruiz E (2011). A study of nanoclay reinforcement of biocomposites made by liquid composite molding. Int J Polym Sci 2011:964193.
35. Kaboorani A, Riedl B, Blanchet P (2013) Ultrasonication technique: A method for dispersing nanoclay in wood adhesives. J Nanomater 2013: 341897.
36. Lam C-K, Lau K-T, Cheung H-Y, Ling H-Y (2005) Effect of ultrasound sonication in nanoclay clusters of nanoclay/epoxy composites. Mater Lett 59(11): 1369–1372.
37. Dean K, Krstina J, Tian W, Varley RJ (2007) Effect of ultrasonic dispersion methods on thermal and mechanical properties of organoclay epoxy nanocomposites.Macromol Mater Eng 292(4): 415–427.
38. Liu Z, Erhan SZ, Xu J (2005) Preparation, characterisation and mechanical properties of epoxidised soybean oil/clay nanocomposites. Polymer 46(23):10119–10127.
39. Roy RK (1990) A primer on the Taguchi method. Van Nostrand Reinhold: New York.
40. Salam H, Dong Y, Davies IJ, Pramanik A (2018) Optimization of material formulation and processing parameters in relation to mechanical properties of bioepoxy/clay nanocomposites using Taguchi design of experiments" J Appl. Polym. Sci. 135(5): 45769.
41. Cravotto G, Cintas P (2011) Introduction to sonochemistry: A historical and conceptual overview. In: Chen D, Sharma SK, Mudhoo A (Eds) Handbook on Applications of Ultrasound: Sonochemistry for Sustainability. CRC/Taylor and Francis: Boca Raton, 23–40.
42. Belavendram N (1995) Quality by design. Prentice Hall International: London.
43. Miyagawa H, Mohanty AK, Misra M, Drzal LT (2004) Thermo-physical and impact properties of epoxy containing epoxidised linseed oil, 2. Macromol Mater Eng 289(7): 636–641.
44. Patel MB, Patel RG, Patel VS (1989) Effects of reactive diluent diepoxidised cardanol and epoxy fortifier on curing kinetics of epoxy resin. J Therm Anal 35(1): 47–57.
45. Ahmed S, Jones FR (1990). A review of particulate reinforcement theories for polymer composites. J Mater Sci 25(12): 4933–4942.
46. Fu S-Y, Feng X-Q, Lauke B, Mai, Y.-W. (2008). Effects of particle size, particle/matrix interface adhesion and particle loading on mechanical properties of particulate–polymer composites. Compos Part B 39(6): 933–961.
47. Miyagawa H, Rich MJ, Drzal LT (2004) Amine-cured epoxy/clay nanocomposites. I. Processing and chemical characterisation. J Polym Sci Part B: Polym Phys 42(23): 4384–4390.
48. Singha S, Thomas M J (2008) Reduction of permittivity in epoxy nanocomposites at low nano-filler loadings. Paper presented at the 2008 Annual Report Conference on Electrical Insulation and Dielectric Phenomena, Quebec, Canada, 2008.
49. Sahoo SK, Mohanty S, Nayak SK (2015) Toughened bio-based epoxy blend network modified with transesterified epoxidised soybean oil: Synthesis and characterisation. RSC Adv 5(18):13674-13691.
50. Nikolic G, Zlatkovic S, Cakic M, Cakic S, Lacnjevac C, Rajic Z (2010) Fast Fourier transform IR characterisation of epoxy GY systems crosslinked with aliphatic and cycloaliphatic EH polyamine adducts. Sensors 10(1):684–696.

51. Zhu L, Wool RP (2006) Nanoclay reinforced bio-based elastomers: Synthesis and characterisation. Polymer 47(24): 8106–8115.
52. Chivrac F, Pollet E, Dole P, Avérous L (2010) Starch-based nano-biocomposites: Plasticiser impact on the montmorillonite exfoliation process. Carbohydr Polym 79(4): 941–947.

Chapter 5
Morphological Structures of Bioepoxy/Clay Nanocomposites with Optimum Material Formulation

5.1 FTIR Analysis

The possible chemical reactions taking place to synthesise bioepoxy/clay nanocomposites with different curing agents were shown in Figs. 5.1 and 5.2. The possible crosslinking reaction of bisphenol A diglycidyl ether (DGEBA) and epoxidised soybean oil (ESO) with sophorone diamine (IPDA) was demonstrated in Fig. 5.1. The crosslinking reaction of epoxy with diamine curing agent was achieved through the adhesion and condensation steps. The curing process was carried out by treating epoxy groups (i.e. phenyl glycidyl ether) with a primary amine to form a secondary amine (reactions a1 and b1) and the further secondary amine with another phenyl glycidyl ether to form a tertiary amine (reactions a2 and b2). Another possible crosslinking reaction is the homo-polymerisation and etherification between epoxy molecules with hydroxyl groups (reactions a3 and b3).

On the other hand, the possible chemical reaction of DGEBA and ESO with methyltetrahydrophtalic anhydride (MTHPA) was illustrated in Fig. 5.2. The reaction was initiated by opening the ring groups of MTHPA with the benzyltrietylammonium chloride catalyst for zwitterions. The further possible reaction was a chemical reaction of zwitterions with the phenyl glycidyl ether groups leading to the formation of alkoxide. Another possible reaction is the propagation of alkoxide with MTHPA to form monoester and further react with other phenyl glycidyl ether groups in order to form diester via esterification mechanism.

With respect to DGEBA/ESO blends, the possible chemical reaction took place in bioepoxy blending with IPDA and MTHPA, which was basically similar to the chemical reaction of epoxy groups with the same curing agents, as previously mentioned in Figs. 5.1 and 5.2. In an IPDA curing agent system, the possible chemical reaction took place by joining epoxy side groups of DGEBA and ESO with primary amine side groups, which was followed by the homo-polymerisation between phenyl glycidyl ether groups of DGEBA and ESO, as well as the etherification of hydroxyl groups with other phenyl glycidyl ether groups. However, with the functionalisation of MTHPA, the possible chemical reaction would occur by the initiation of MTHPA

H. Salam and Y. Dong, *Bioepoxy/Clay Nanocomposites*,
https://doi.org/10.1007/978-981-16-7297-2_5

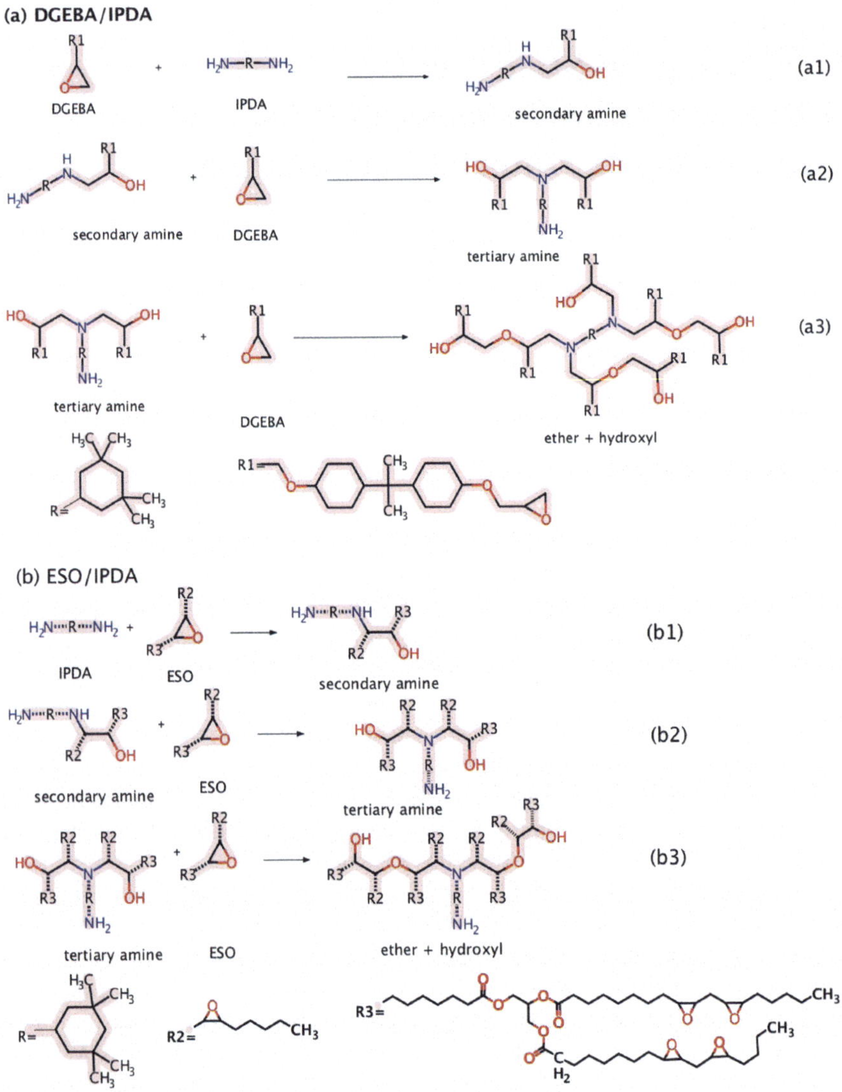

Fig. 5.1 Possible chemical reaction mechanisms: **a** DGEBA and **b** ESO cured by IPDA [1]

at an elevated temperature using a thermal latent catalyst to activate a dipolar ion of anhydride side groups. This reaction was followed by nucleophiles (SN2) reaction of zwitterions with epoxy groups for both epoxy and ESO leading to an alkoxide. The reaction was then followed by the crosslinking reaction through the esterification of alkoxide with an anhydride curing agent to form carboxylate anion.

Important characteristic peaks of epoxy groups in DGEBA resin and ESO can be confirmed by using FTIR analysis, as shown in Fig. 5.3. The presence of stretching

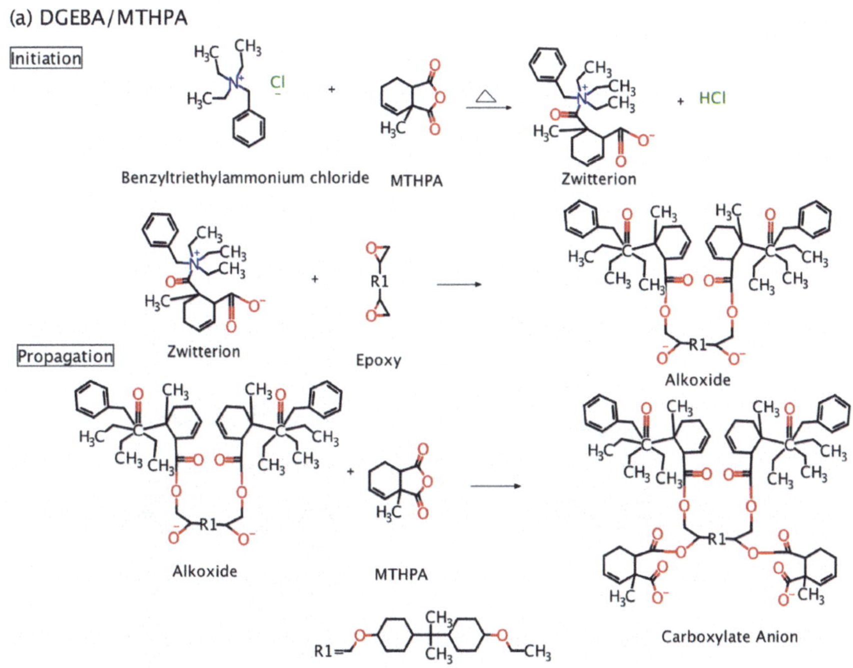

Fig. 5.2 Possible chemical reaction mechanisms: **a** DGEBA and **b** ESO cured by MTHPA [2]

C–O–C of oxirane groups and stretching C-O of oxirane for epoxy group peaks appeared at low wave numbers of 828 and 915 cm^{-1}, respectively. However, those characteristic peaks of epoxy groups were detected at 827 and 842 cm^{-1} [3]. Additionally, other characteristic peaks for DGEBA and ESO were absorbed at 1036 cm^{-1} for stretching C–O–C of ethers, 1182 cm^{-1} for C–C-O-C stretching, 1508 cm^{-1} for stretching C–C of aromatic and 1608 cm^{-1} for stretching C = C of aromatic rings. Nonetheless, the peak at 1727 cm^{-1} corresponded to the stretching of –C = O in the carbonyl group and in range of 2965–2873 cm^{-1} for the stretching of C-H of CH_2 and C-H aromatic and aliphatic. Other characteristic peaks of epoxy groups were absorbed at 3057 cm^{-1} for DGEBA and at ~ 2900 cm^{-1} for ESO, representing the stretching of C-H of oxirane ring. In addition, the FTIR characteristic peaks of Cloisite 15 clays were also presented in Figs. 5.3 and 5.4. The FTIR spectra of Cloisite 15 revealed the presence of characteristic peaks for the O–H stretching at 3627 cm^{-1} and the broadened peak of Si–O-Si stretching at 1004 cm^{-1}. In addition, organomodified molecules of alkylammoinum were absorbed at 2924 and 2842 cm^{-1} for the asymmetry and symmetry of C-H vibration, respectively, as well as C-H bending of methylene groups at 1475 cm^{-1} [4].

With respect to the incorporation of IPDA, as shown in Fig. 5.3a, FTIR spectra presented the decreasing intensity at the peaks of ~ 828, 915 and ~ 3057 cm^{-1}, corresponding to stretching C–O–C, stretching C-O of oxirane for epoxy groups

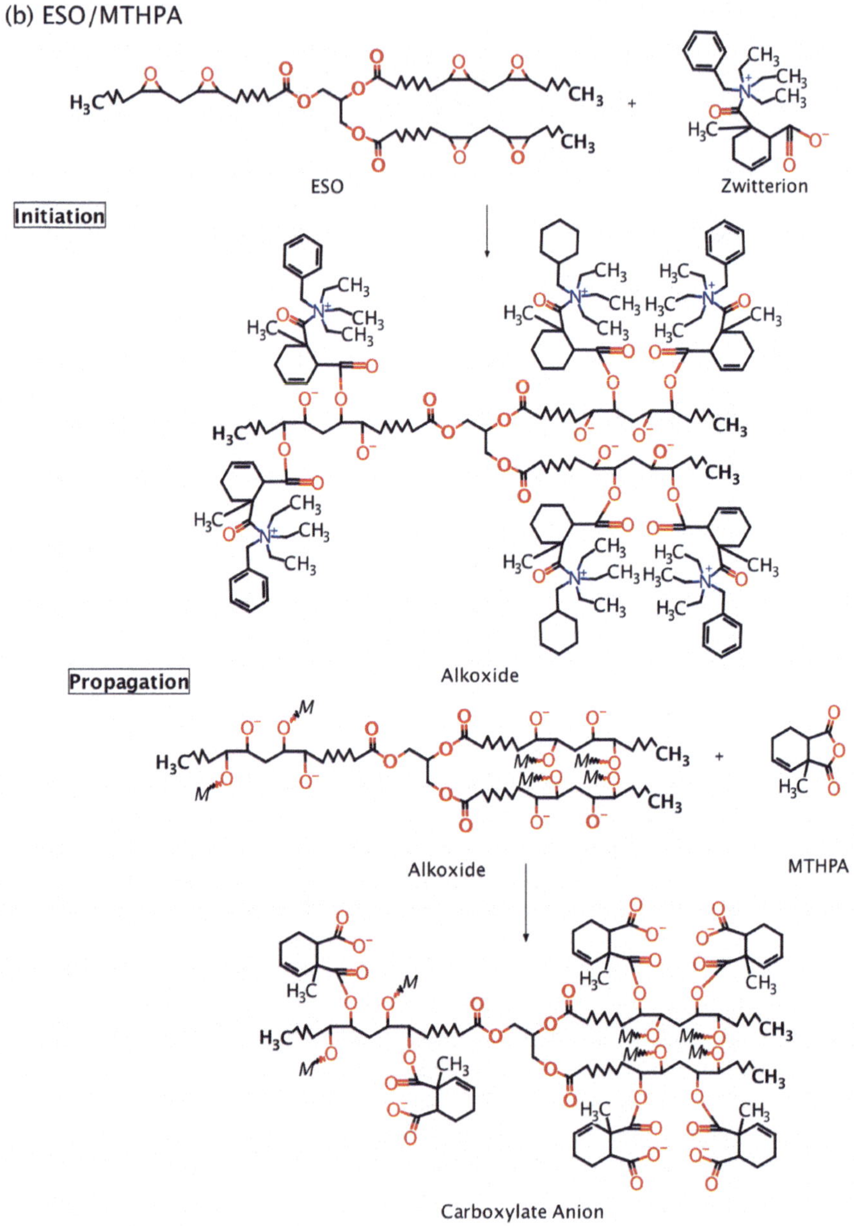

Fig. 5.2 (continued)

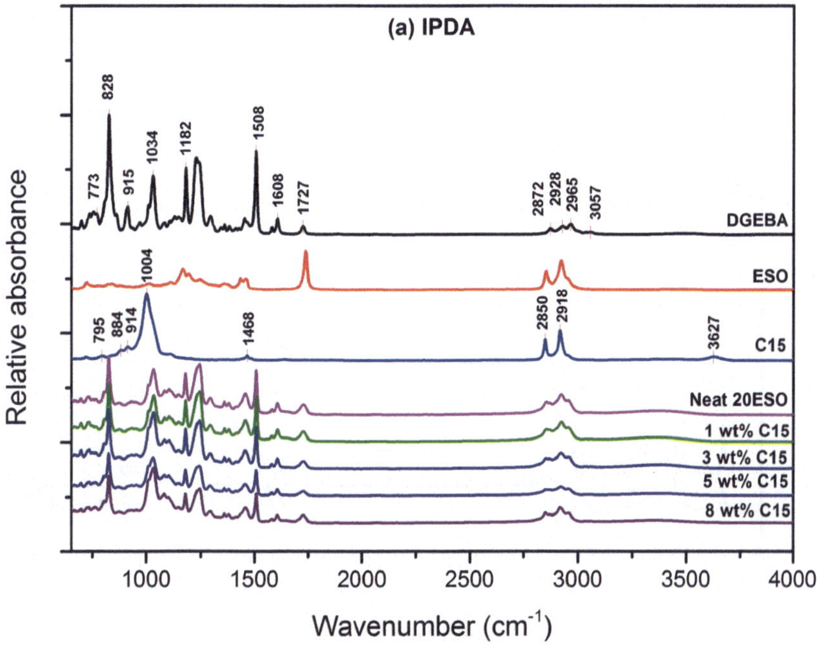

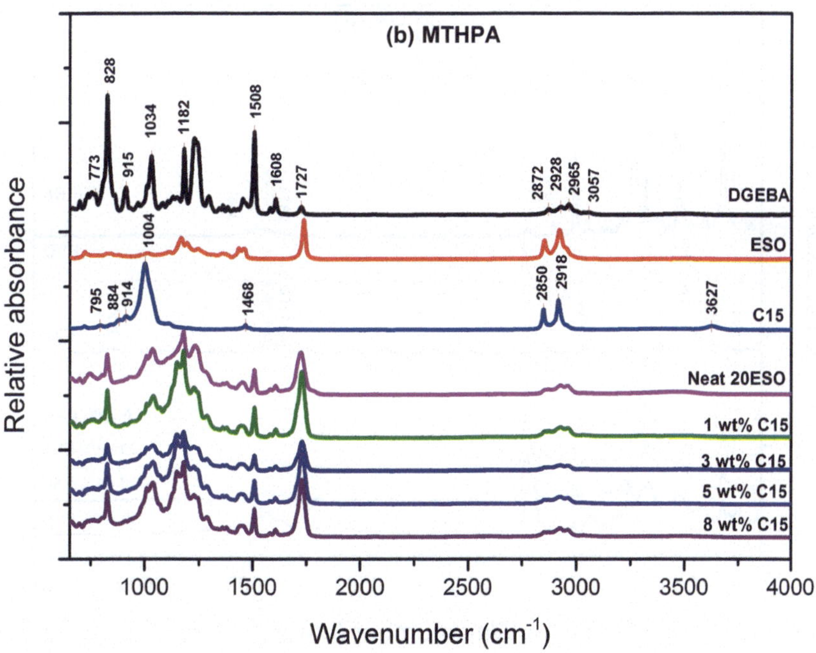

Fig. 5.3 FTIR spectra of bioepoxy/clay nanocomposites at the ESO content of 20 wt% with different clay contents cured by **a** IPDA and **b** MTHPA. The letter C15 is referred to as Cloisite 15 clays [5]

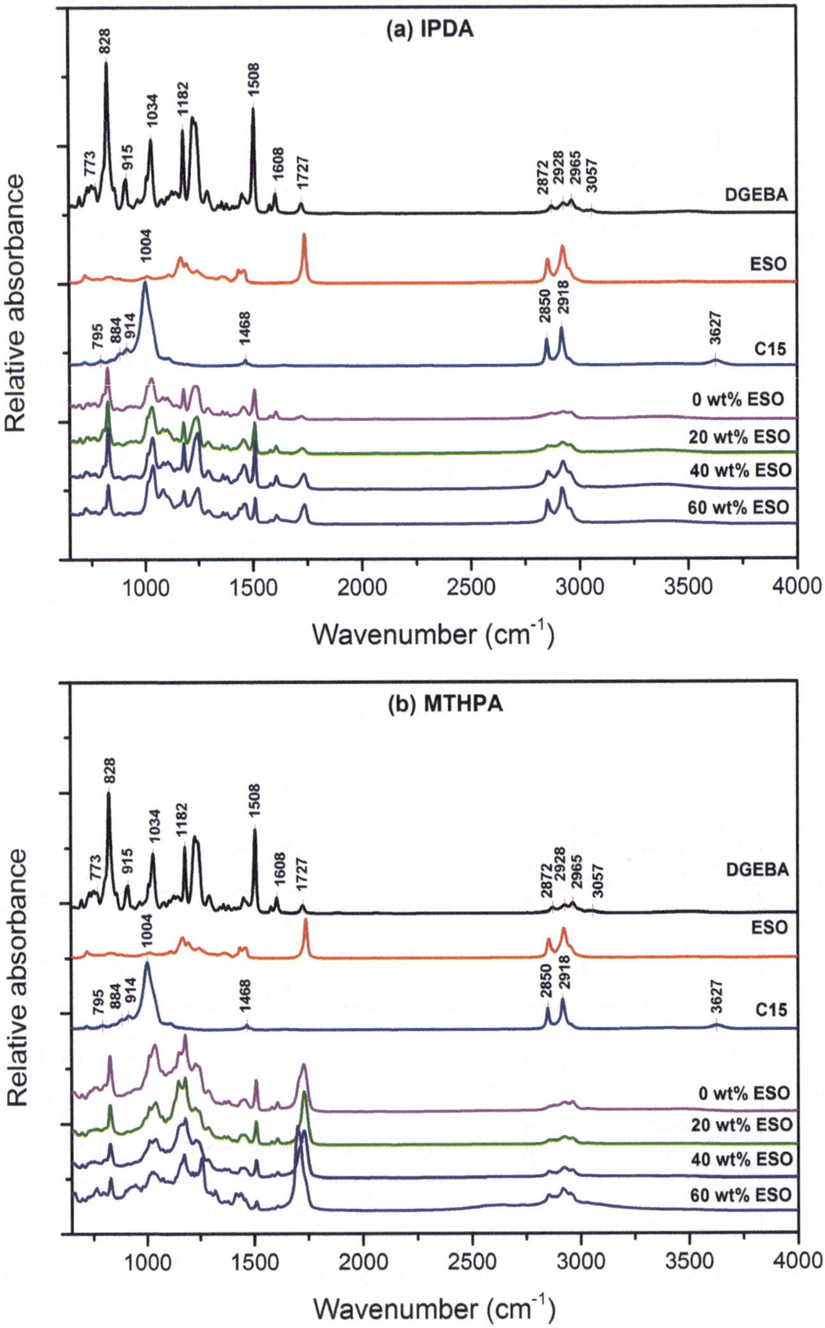

Fig. 5.4 FTIR spectra of bioepoxy/clay nanocomposites reinforced with 5 wt% Cloisite 15 clays with different ESO contents cured by **a** IPDA and **b** MTHPA. The letter C15 is referred to as Cloisite 15 clays [5]

and C-H of oxirane rings accordingly. Nonetheless, a broad spectrum in the range of 3200–3500 cm^{-1} was assigned to the O–H band. These features indicated that the crosslinking reaction between epoxy groups of matrices and amine curing agent occurred to form hydroxyl groups [6]. In addition, the incorporation of Cloisite 15 clays in bionanocomposites systems cured by IPDA demonstrated a non-significant peak change due to the addition of clay fillers. A small peak of Si–O-Si stretching at 1004 cm^{-1} was hidden under the shoulder of a C–O–C stretch band of epoxy groups at 1034 cm^{-1}. Nonetheless, it became visible along with increasing the clay content from 1 to 8 wt% Cloisite 15 clays. On the other hand, the effect of clay content and ESO content on bionanocomposite structures was detected from FTIR spectra in Figs. 5.3 and 5.4. In an IPDA curing system, the incorporation of clay fillers into bioepoxy matrices resulted in a decrease in the peak of carbonyl groups at ~ 1727 cm^{-1}. This finding suggested that the interaction between epoxy groups and clays was developed in the clay platelet layers, in good accordance with previous literature [7].

On the other hand, the effect of functionalisation of MTHPA on bionanocomposite structures was presented in Fig. 5.3b. Similar to bionanocomposites systems cured by IPDA, the incorporation of MTHPA reduced the peak absorbance at the band of 828 cm^{-1} corresponding to the stretching C–O–C with the peak absence taking place at 915 cm^{-1}, which was assigned to the stretching C-O of oxirane for epoxy groups. On the contrary, the peak area associated with the -C = O vibration of ester at ~ 1727 cm^{-1} increased for neat bioepoxy and bionanocomposites with 20% ESO at all clay contents. Nonetheless, a new peak at the wave number of ~ 1160 cm^{-1} was assigned to the C-O vibration or crosslinked ester and ether linkages, which appeared in the spectra of cured bionanocomposites. These features indicated that the crosslinking reaction of epoxy groups with MTHPA occurred involving the polyesterification reactions in bionanocomposites system cured by MTHPA [8]. Additionally, the inclusion of Cloisite 15 clays to bionanocomposite systems cured by MTHPA gave rise to an overlap peak characteristic of Cloisite 15 clays at 1004 cm^{-1} assigned to the band stretching of Si–O-Si along with C–O–C stretching of epoxy groups taking place at 1034 cm^{-1}. Other peak characteristics of Cloisite 15 clays did not demonstrates significant changes due to low peak absorbance.

With respect to bionanocomposites at different ESO contents, Fig. 5.4 presented FTIR spectra of bionanocomposite systems cured by both IPDA and MTHPA. As mention previously, the incorporation of IPDA and MTHPA into bionanocomposites systems reduced the peaks of epoxy groups at 828 and 915 cm^{-1} in accordance with epoxy groups of C–O–C stretching and C-O stretching of oxirane, respectively. Nonetheless, the overlap peak characteristics of Cloisite 15 clays and epoxy groups were observed at 1004 and 1034 cm^{-1}, which were assigned to stretching of Si–O-Si and C–O–C, respectively. In addition, with increasing the ESO content from 0 to 60 wt%, FTIR spectra of bionanocomposites cured by IPDA (Fig. 5.4a) demonstrated an increasing peak absorbance at four different peak characteristics of ESO, which was referred to as 1460, 1740 and 2854–2928 cm^{-1}, and corresponded to $-CH_2$ bending, -C = O vibration of ester and C-H stretching of CH_2 and C-H aromatic and aliphatic, respectively. Nonetheless, the peak of C = O vibration of ester demonstrated a slight peak shift from a wave-number range from 1727 to 1740 cm^{-1} with increasing the ESO content from 0 to 60 wt%. The enhanced peak absorbance and shifted peak frequency of ESO characteristic indicated the presence of unreacted ESO with increasing the ESO content. These findings were discussed in the further section of SEM analysis results for bionanocomposites cured by IPDA at different ESO contents. Nonetheless, in bionanocoposite systems cured by IPDA, The spectra in a range of 3200–3500 cm^{-1} corresponding to O–H band were broadened, indicating the crosslinking reaction of epoxy groups and amine curing agents [6].

On the other hand, in bionanocomposites systems cured by MTHPA at different ESO contents (Fig. 5.4b), FTIR spectra presented a reduction of aromatic stretching C–C and stretching C = C of aromatic rings at 1509 and 1608 cm^{-1}, respectively. On the contrary, an increasing peak absorbance at approximately 1161 and 2928 cm^{-1} were observed in relation to C-O stretching ester band and C-H stretching of CH_2 and C-H aromatic and aliphatic with increasing the ESO content. In addition, the peak characteristic absorbance for -C = O vibration of ester at 1727 cm^{-1} appeared to be enhanced for both conventional nanocomposites and bionanocomposites at different ESO contents. Nonetheless, at the ESO content of 60 wt%, the peak characteristic of C = O vibration of ester shifted to a lower wave number of 1699 cm^{-1}. This feature indicated that the resonance, as well as conjugation lowered the stretching of carbonyl groups (-C = O) [9]. Such results associated with wave number reduction revealed a decrease in the degree of polymerisation for bioepoxy blends owing to the addition of ESO in possession of a rubber-like material behaviour.

5.2 XRD Analysis

5.2.1 Effect of Clay Content

XRD patterns offer an initial evaluation of clay dispersion status in bioepoxy matrices. The effect of clay content and functionalisation of curing agents on morphological structures of bionanocomposites was investigated. Figure 5.5 showed specific diffractograms of bioepoxy/clay nanocomposites at a fixed ESO content of 20 wt% with different clay contents and curing agent types. The XRD pattern for Cloisite 15 clays presented three peaks at low 2θ ranges from 1 to 10°. The first (001) peak appeared at $2\theta = 2.55°$ corresponding to interlay spacing $d_{001} = 3.46$ nm. The second (002) peak characteristic was located at $2\theta = 4.76°$ ($d_{002} = 1.85$ nm) and the third peak (003) occurred at $2\theta = 7.24°$ ($d_{003} = 1.22$ nm). These peaks revealed a wide range of interlayer spacing of clay platelet layers in Cloisite 15 clays with different *d*-spacing values.

The XRD diffractograms in Fig. 5.5a were presented for bionanocomposites with 20 wt% ESO cured by IPDA at different clay contents (i.e. 1 to 8 wt%). The addition of 1 wt% Cloisite15 clays into such a bionanocomposite system demonstrated (002) peak shift took place at $2\theta = 4.49°$, corresponding to the interlayer spacing d_{002} = 1.97 nm. Similarly, one peak representing (002) peak shift evidently appeared

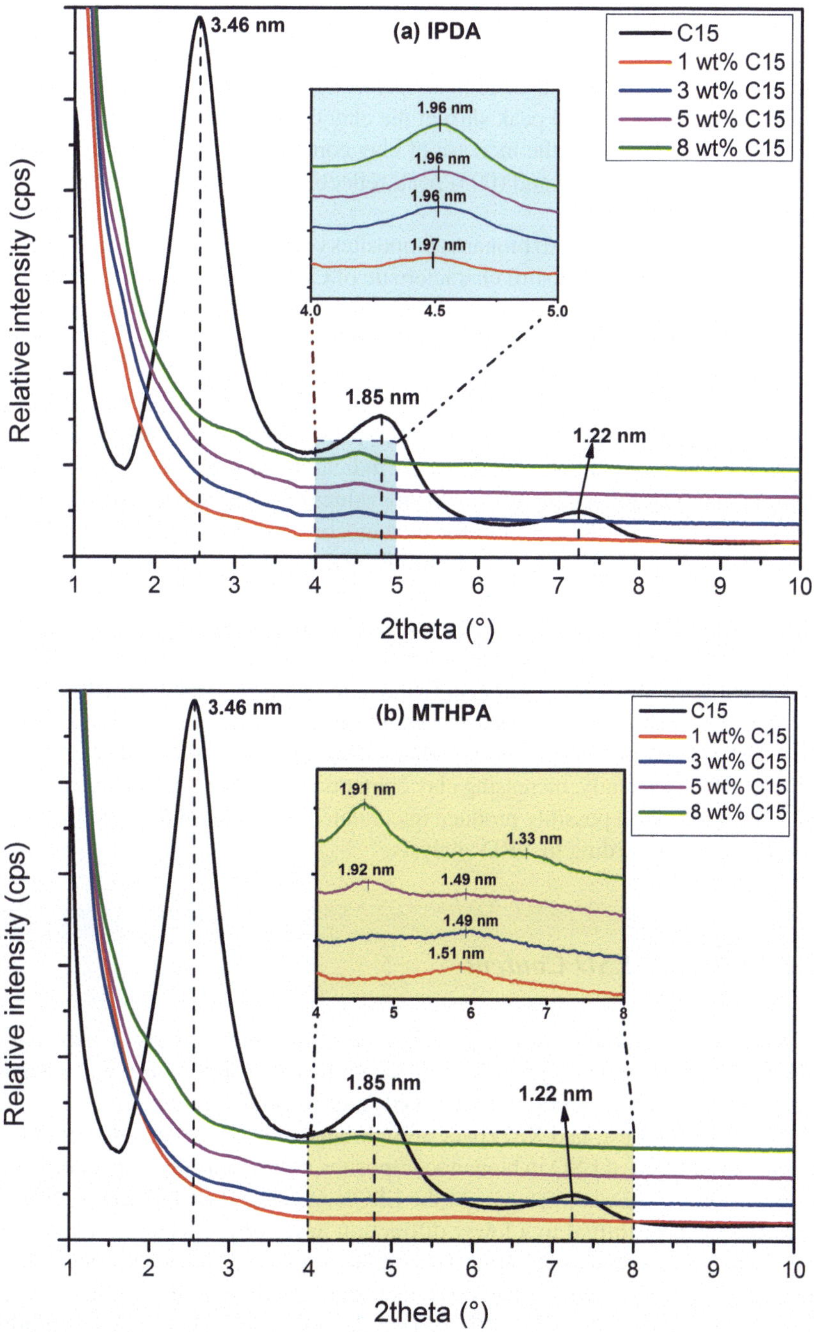

Fig. 5.5 XRD patterns of bioepoxy/clay nanocomposites (ESO content: 20 wt%) at different Cloisite 15 clay contents cured by **a** IPDA and **b** MTHPA. The letter C15 is referred to as Cloisite 15 clays [5]

in a bionanocomposite system cured by IPDA with the inclusion of 3 to 8 wt% Cloisite15 clays. The (002) peak shifted to a lower angle of diffraction, which was identified at 2θ ≈ 4.52° and resulted in an increase in interlayer spacing (d_{002}) about 0.11 nm. Nonetheless, (002) peak shift at the clay contents of 3 to 8 wt% yielded a linear relationship between the increase of clay content and the area increase under (002) peak. However, (001) and (003) peak reflections were also manifested in the measured 2θ range.

Meanwhile, with respect to bionanocomposites cured by MTHPA (Fig. 5.5b), the XRD pattern showed a peak shift characteristic of Cloisite 15 clays. (002) peak shift appeared with the addition of 5 and 8 wt% Cloisite 15 clays in such bionanocomposites. The *d-spacing* values of (002) reflection increased for about 0.07 and 0.06 nm only corresponding to bionanocomposites reinforced with 5 and 8 wt% Cloisite 15 clays, respectively, as opposed to 1.85 nm for the (002) reflection of raw Cloisite 15 clays. Moreover, Fig. 5.5b also exhibited (003) peak shift to a lower diffraction angle compared to organomodified clay powders, indicating an increase in the interlayer spacing at the (003) reflection. The *d-spacing* values of (003) peaks were determined to be 1.51, 1.49, 1.49 and 1.33 nm for bionanocomposites reinforced with 1, 3, 5 and 8 wt% Cloisite 15 clays. This again confirmed the existence of intercalated clay structures within bionanocomposites reinforced with Cloisite 15 clays when cured by MTHPA. Moreover, peak-disappearance and peak-broadening phenomena with respect to (001) reflection also happened in bionanocomposites cured by MTHPA at different clay contents, which was also found in bionanocomposites cured by IPDA. The peak disappearance suggested the occurrence of local exfoliated clay structures or disordered orientation of clay platelets/aggregates in polymer matrices for (001) peak. Consequently, increasing clay contents from 1 to 8 wt% in bionanocomposite materials could possibly produce intercalated structures at (002) reflection of Cloisite15 clays according to XRD analysis.

5.2.2 Effect of ESO Content

The XRD analysis for clay structures in bionanocomposites was also conducted at different ESO contents in the range of 0–60 wt% when reinforced with 5 wt% Cloisite 15 clays and cured by both IPDA and MTHPA, as illustrated in Fig. 5.6.

Figure 5.6a demonstrated XRD patterns of bionanocomposites cured by IPDA. In general, the addition of ESO in bionanocomposites influenced clay dispersion status and their morphological structures. At the ESO content of 0 wt%, XRD peak (001) of Cloisite15 clays shifted to a lower diffraction angle of 2θ, which resulted in an increase in d_{001} from 3.46 to 3.79 nm. This finding suggested typical clay intercalated structures taking place within bioepoxy matrices. In addition, the inclusion of 20 wt% ESO did not exhibit any (001) peak reflection. Meanwhile, the substitution of conventional epoxy with 40 and 60 wt% ESO demonstrated (001) peak shift to a higher diffraction angle at $2\theta \approx 2.85$ and 2.95°, corresponding to d_{001}= 3.10 and 3.01 nm, respectively. This d_{001} peak reduction indicated a collapse of clay

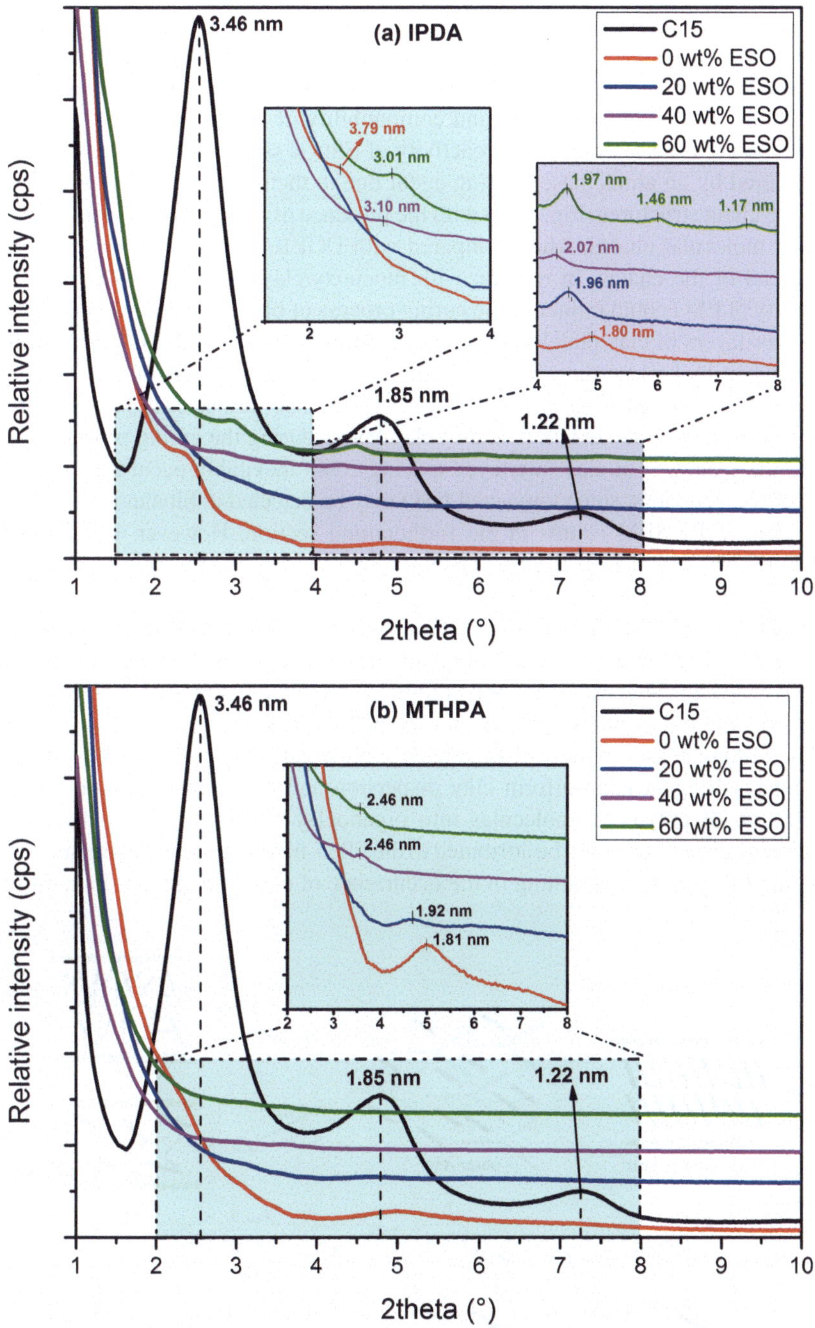

Fig. 5.6 XRD patterns of bioepoxy/clay nanocomposites reinforced with 5 wt% Cloisite 15 clays at different ESO contents cured by **a** IPDA and **b** MTHPA. The letter C15 is referred to as Cloisite 15 clays [5]

structures when using the higher ESO contents of 40 and 60 wt%. The highest ESO content of 60 wt% yielded higher intergallery shrinkage for about 0.45 nm as compared to 0.36 nm with 40 wt% ESO. This shrinkage might be associated with two factors, namely the reactivity and compatibility of epoxy matrices and curing agent, as well as clay content. The reactivity of ESO is smaller than that of DGEBA when cured by an amine based curing agent due to their chemical structures. ESO has long chain structures (Fig. 5.1b) with the presence of epoxy groups in the middle of their molecular chains, when compared with DGEBA containing epoxy groups at the end of the chains. In relation with bioepoxy/clay nanocomposites, the less reactivity of ESO could influence the curing process of blended matrices in the inner and outer-layers of clay platelets. The crosslinking reaction of DGEBA with amine curing agent took place more rapidly in the inner and outer-layers of clay platelets with only the limited ESO reaction. Meanwhile, some unreacted ESO molecules would be pushed to the external side of clay layers during the curing process. As a result, the reduction of clay interlayer spacing could inevitably occur illustrated in Fig. 5.7. Nonetheless, some unreacted ESO may be detected in bionanocomposites, as evidenced by SEM results in the forthcoming section. However, in related to molecular weights of DGEBA and ESO, the higher molecular weight might make it difficult to penetrate into interlayer areas of clay layers. The molecular weight of DGEBA at 340.42 g/mol is lower than that of ESO at 975.40 g/mol according to Table 2.1. Furthermore, at the ESO content of 60 wt%, the highest reduction in interlayer spacing suggested that only a limited amount of epoxy molecules enabled to penetrate into clay interlayer areas for the polymerisation.

The second reason could be associated with organoclay type and higher clay content resulting in non-uniform clay dispersion and consequently hindering the penetration of polymeric molecules into organoclay interlayer areas. The worse organoclay dispersion could be attributed to the filler–filler interaction of organoclays at higher filler contents resulting in the occurrence of clay aggregation. In addition,

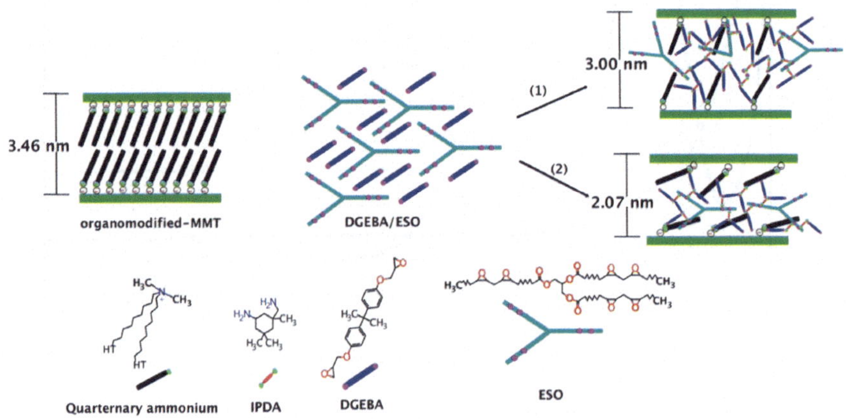

Fig. 5.7 Schematic diagrams for the possible shrinkage process in clay interlayer areas [2]

Cloisite 15 clays as commercial materials with quaternary onium ions tended to produce clay tactoid structures of typical intercalated nanocomposites, as confirmed by Lan et al. [10]. During the curing process, quaternary onium ions are exchanged by epoxy resin monomers from clay interlayers. The number of exchanged onium ions would be influenced by the availability of layer negative charge for the starting clays, thus influencing the *d*-spacing.

Nonetheless, as illustrated in Fig. 5.6a, XRD patterns of bionanocomposites cured by IPDA with different ESO contents also revealed a peak characteristic at (002) reflection. The bionanocomposites with 0 wt% ESO possessed a small peak of (002) reflection at $2\theta \approx 4.89°$ in relation to the *d-spacing* value of 1.80 nm. This (002) reflection demonstrated a decreasing interlayer spacing of (002) reflection about 0.05 nm, owing to the shrinkage of interlayer spacing. However, the increase in the ESO content from 20 to 60 wt% exhibited an increasing d_{002} values of Cloisite 15 clays to approximately 1.96, 2.07 and 1.97 nm, respectively. Such an enhancement implied that intercalated structure was detected at (002) reflection along with increasing the ESO content. Additionally, bionanocomposites with 60 wt% ESO also possessed a small peak close to (003) reflection for Cloisite 15 clays at $2\theta \approx 6.05$ and 7.53°, respectively. Those peaks signified that clay intercalation also existed at (003) peak characteristic by an increase in d_{003} up to approximately 1.46 nm, as well as a decrease in d_{003} of Cloisite 15 clays to 1.17 nm.

On the other hand, Fig. 5.6b presented XRD patterns of bionanocomposites reinforced with 5 wt% Cloisite 15 clays, which only occurred at different ESO contents cured by MTHPA. In conventional nanocomposites (i.e. 0 wt% ESO), the XRD pattern showed (002) peak shift into higher diffraction angle, suggesting clay compacted structures in relation to (002) reflection. The d_{002} of bionanocomposites without ESO slightly decreased to approximately 1.81 nm. However, increasing the ESO content from 20 to 60 wt% demonstrated the peak shift of (002) reflection to the lower diffraction angle with a clear sign of clay intercalation. The interlayer spacings of (002) reflection increased by about 4, 33 and 33% with respect to $d_{002} = 1.85$ nm for Cloisite 15 clays, which could be referred to as bionanocomposites with 20, 40 and 60 wt% ESO, respectively. Meanwhile, (001) and (003) peak characteristics of Cloisite 15 clays did not exhibit any peak reflection in 2θ range from 1 to 10°. This peak disappearance might imply that (001) and (003) reflections resulted in the random orientation of clay dispersed structures leading to a broadened XRD peak with lower intensity value, which, however, could not be identified by XRD analysis. Consequently, those findings in Fig. 5.2 suggested that increasing the ESO content was expected to enhance interlayer spacing at (002) reflection in order to generate intercalated clay structures.

On the other hand, the functionalisation of curing agents presented different clay dispersion statuses in bionanocomposites. As seen in Figs. 5.5 and 5.6, bionanocomposites cured by IPDA demonstrated a peak shift of (002) reflection to lower 2θ angles, which appeared to be less pronounced when MTHPA was used. Various curing agents demonstrated different reactivities in the curing process for the intergallery and outergallery of clay platelet layers. In particular, the reaction rate of bioepoxy in this study relied on the balance between curing agent type and reaction rate since a variety of curing agents underwent different curing processes, thus easily influencing the reaction rate. Bionanocomposites cured by MTHPA demonstrated a peak-broadening phenomenon in the XRD pattern, which could be influenced by the functionalisation in the curing process at elevated temperatures. Higher curing temperature used for the fabrication of bionanocomposites can influence the reactivity for the initiation of curing process, as well as improve the delamination of clay platelets [11]. The use of higher curing temperature enabled to reduce the viscosity leading to the improvement of molecular mobility for their diffusion into clay interlayers. Additionally, MTHPA belongs to an anhydride curing agent with the low viscosity, longer gel time for 8 h and less hazards when compared with amine curing agents [12–14]. The longer gel time of anhydride curing agent tends to minimise different polymerisation reaction rate in clay intergalleries and outergalleries for the equilibrium. Since XRD analysis is generally used to evaluate intercalated clay structures, TEM observation is essential as a supplementary characterisation technique to visually detect the clay dispersion status especially in terms of exfoliated or aggregated clay structures in the section mentioned below.

5.3 TEM Analysis

Clay dispersion status in bioepoxy matrices is very complex, which is influenced by several factors such as processing parameters, bio-renewable content, curing agent and clay properties [15]. In this section, the effect of clay content, ESO content and curing agent type on clay dispersion status in bioepoxy matrices was discussed. TEM images of bionanocomposites with different clay contents, ESO contents and curing agent types were depicted in Figs. 5.8, 5.9 and 5.10. The dark lines and clouds represented clay platelet layers and clay aggregates, respectively. In contrast, bright areas were identified as polymer matrices. In bionanocomposite systems cured by IPDA at the ESO content of 20 wt% (Fig. 5.8), organoclay fillers were moderately dispersed in bioepoxy matrices with the aggregate size of 2–6 μm depending on the clay content. In bionanocomposites reinforced with 1 wt% Cloisite 15 clays, clay particles were moderately dispersed with an aggregate size of approximately 2 μm (Fig. 5.8a). In addition, it was also observed that intercalated and exfoliated clay structures occurred in the matrices (Fig. 5.8b). The exfoliated structures were shown to be randomly dispersed with some intercalated structures of clay platelets within bioepoxy matrices. On the other hand, in bionanocomposites reinforced with 3–8 wt% Cloisite 15 clays, dominant clay aggregation was manifested in the images as opposed to intercalated and exfoliated structures. Clay aggregates were exhibited in size of ~2 μm. Those images clearly suggested the poor wettability of clay fillers in bioepoxy matrices.

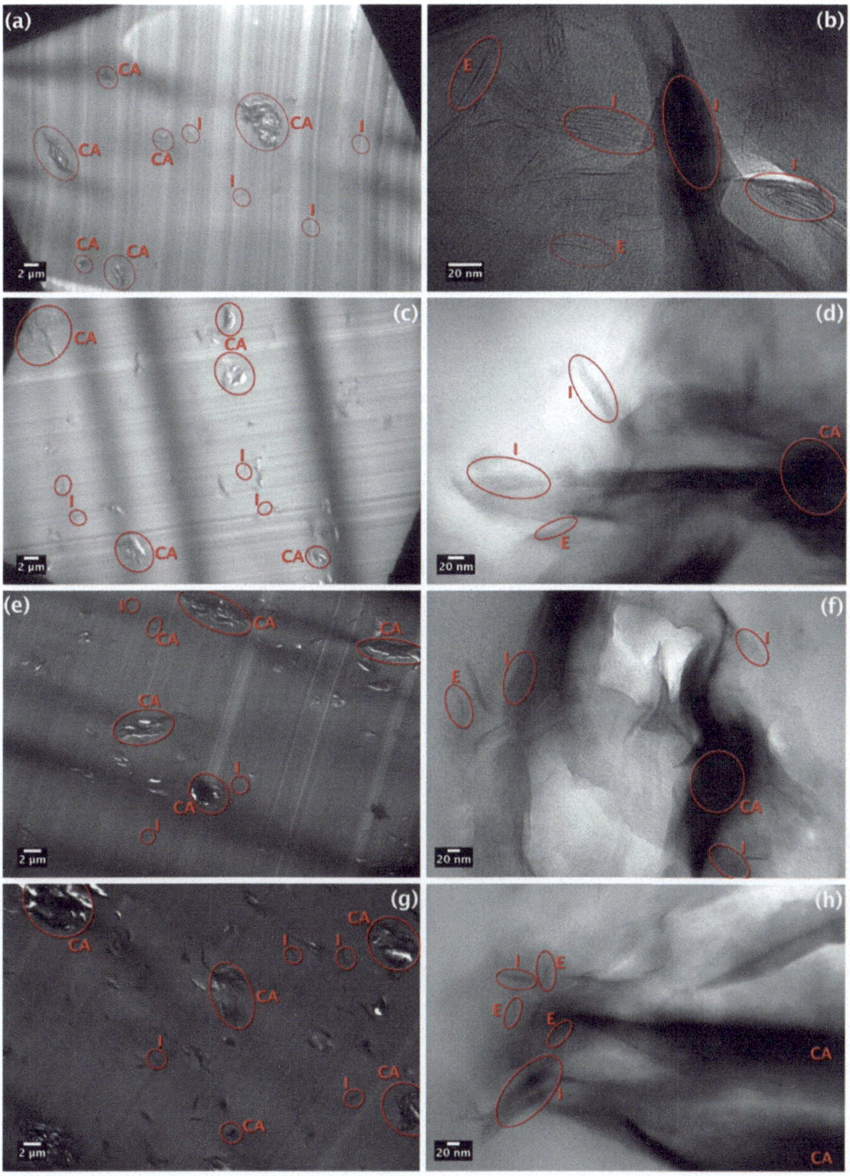

Fig. 5.8 Clay dispersion statuses in bionanocomposites cured by IPDA (ESO content: 20 wt%) at different Cloisite 15 clay contents with various magnifications **a** 1 wt% at 500 × and **b** at 30,000 × ; **c** 3 wt% at 500 × and **d** at 30,000 × ; **e** 5 wt% at 500 × and **f** at 30,000 × , as well as **g** 8 wt% at 500 × and **h** at 30,000 × . The letters CA, I and E represent clay aggregates, intercalated and exfoliated structures, respectively [5]

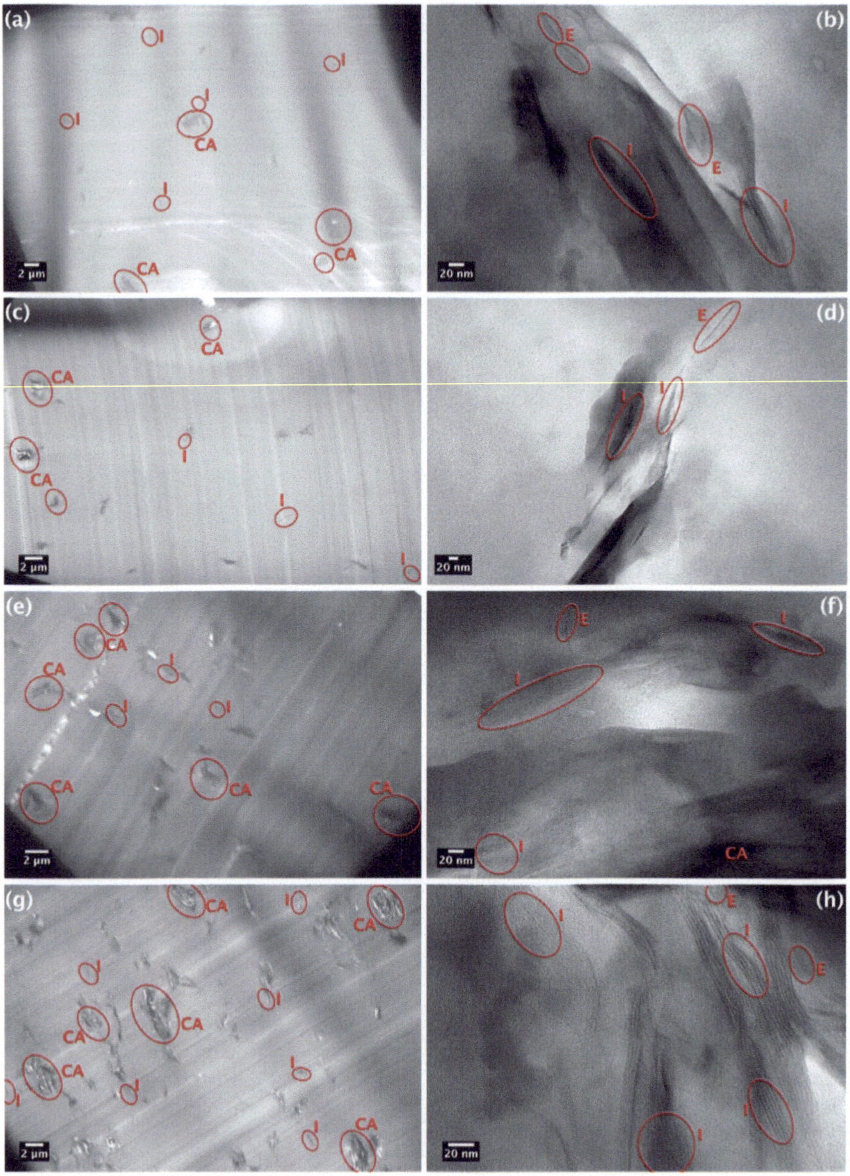

Fig. 5.9 Clay dispersion statuses in bionanocomposites cured by MTHPA at different Cloisite 15 clay contents with various magnifications. **a** 1 wt% at 500 × and **b** at 30,000 × ; **c** 3 wt% at 500 × and **d** at 30,000 × ; **e** 5 wt% at 500 × and **f** at 30,000 × , as well as **g** 8 wt% at 500 × and **h** at 30,000 × . The letters CA, I and E represent clay aggregates, intercalated and exfoliated structures, respectively [5]

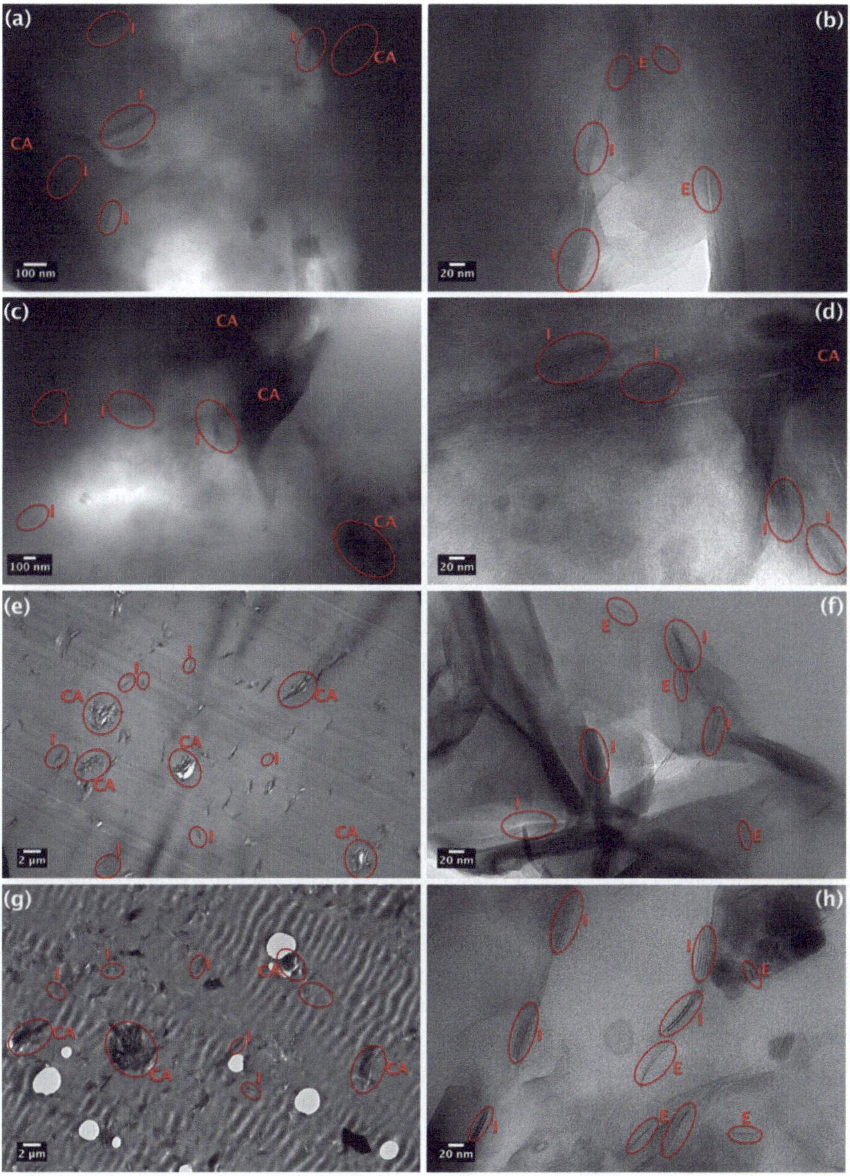

Fig. 5.10 Clay dispersion statuses in bionanocomposites systems at different curing agents and ESO contents reinforced with 5 wt% Cloisite 15 clays at various magnifications. **a** 40 wt% ESO/IPDA at 15,000× and **b** at 30,000 × ; **c** 60 wt% ESO/IPDA at 8000 × and **d** at 30,000 × ; **e** 40 wt% ESO/MTHPA at 500 × and **f** at 30,000 × , as well as **g** 60 wt% ESO/MTHPA at 500 × and **h** at 30,000 × . The letters CA, I and E represent clay aggregates, intercalated and exfoliated structures, respectively [5]

On the other hand, bionanocomposites with 20 wt% ESO cured by MTHPA also demonstrated a mixed clay dispersion status, in which disordered clay orientation with partially intercalated/exfoliated and aggregated clay structures became manifested (Fig. 5.9). In particular, clay aggregate size appeared to be enhanced with increasing the clay content. For instance, at the highest clay content of 8 wt%, clay aggregates were found to be in size of approximately 5 μm, Fig. 5.9g. On the other hand, exfoliated/intercalated clay structures were shown to be disoriented and overlapping between clay layers at different clay contents in range from 1–8 wt%. This TEM finding has further confirmed XRD analysis results, demonstrating broadened XRD patterns, as depicted in Fig. 5.5b. Overall, bionanocomposites cured by IPDA and MTHPA at the fixed ESO content of 20 wt% demonstrated disordered Cloisite 15 clay structures with a combination of partially intercalated/exfoliated and aggregated clays regardless of different clay contents, which was also validated for clays with quaternary onium ion type elsewhere [10].

Furthermore, TEM analysis was also used to observe the effect of ESO content on clay dispersion status in bionanocomposites, shown in Fig. 5.10. As can be seen in Fig. 5.10a–d, clay dispersion status in bionanocomposites cured by IPDA presented aggregated clays with partially intercalated clay platelets within the matrices. This finding supported XRD analysis results that the substitution of DGEBA up to 60 wt% with ESO was likely to produce intercalated structures of clay platelet layers. Conversely, the dispersion status of clay particles in bionanocomposites cured by MTHPA at the ESO contents of 40 and 60 wt% demonstrated moderate clay dispersion in bioepoxy matrices (Fig. 5.10e–h). Increasing the ESO content in bioepoxy components influenced the clay dispersion status in bionanocomposites. Better clay dispersion was evident when the ESO content increased from 20 to 40 wt%. Such a phenomenon could be attributed to the reduced viscosity of bioepoxy blends, which improved the molecular mobility of bioepoxy matrices to delaminate and diffuse into clay interlayers [16, 17]. ESO has a relatively low viscosity of ~ 325 mPa·s as compared to DGEBA at ~11.000 mPa·s (Table 2.1). The increase in ESO content could reduce the viscosity of DGEBA and enhance the mobility of molecular chains. With a low viscosity for bioepoxy blends, the molecules of bioepoxy matrices could diffuse into clay interlayers to form typical intercalation structures, as well as delaminate clay layers starting from the external areas leading to exfoliated structures.

Better clay dispersion (Fig. 5.10e) has been detected at the ESO content of 40 wt% when compared with that at 60 wt% (Fig. 5.10g). Clay aggregates were shown to exist in size of approximately 0.5 μm for bionanocomposites with 40 wt% ESO. Nonetheless, an increase in clay aggregate size (about 1–3 μm) was observed for those with 60 wt% ESO. In particular, at the high magnification illustrated in Fig. 5.10f and h, intercalated clay structures appeared to be more dominant as compared with exfoliated and aggregated structures of bionanocomposites with 40 and 60 wt% ESO. These findings suggested that increasing ESO content was likely to produce bionanocomposites mainly with a mix of tactoids and typically intercalated clay structures.

5.4 SEM Analysis

Fracture surfaces of neat bioepoxy and bionanocomposites with different clay contents, ESO contents and curing agents were examined by SEM analysis, and the relevant micrographs were displayed in Figs. 5.11, 5.12, 5.13 and 5.14, in which letters CA and MC represented clay aggregates and microcracks, respectively, and an arrow denoted the crack due to the applied load. Figures 5.11 showed fracture surfaces of bioepoxy matrices with different curing agents in bionanocomposites. It could be seen that smooth surface finish with a sign of crack propagation was clearly shown due to the applied load during the tests. Moreover, the smooth and featureless surface morphology indicated the characteristic of amorphous brittle materials of bioepoxy with stream marks adjacent to initial cracks. The functionalisation of various curing agents presented different fracture surface characteristics. As seen in Fig. 5.6, the fracture surface of bioepoxy cured by IPDA (Fig. 5.11a) presented slightly textured fracture surface as compared with those cured by MTHPA (Fig. 5.11b) with a river-like fracture surface pattern.

The effects of different clay content on fracture surfaces were depicted in Figs. 5.12 and 5.13, which was corresponding to bionanocomposites cured by IPDA and MTHPA, respectively. An increase in clay content within bioepoxy matrices presented moderate clay dispersion for both bionancomposites with different curing agents (i.e. IPDA and MTHPA). As shown in Fig. 5.12, in addition to the brittle surface characteristic, fracture surfaces of bionanocomposites exhibited textured surfaces and rough surface morphology owing to the existence of embedded clays within bioepoxy matrices. The addition of 1 wt% Cloisite 15 clays presented microcracks as a typical fracture surface characteristic, resulting from weak interfacial bonding in the localised areas when compared with other areas after the applied load during the testing. Nonetheless, Fig. 5.12a also presented the existence of aggregated clay particles in fracture surfaces with the aggregate size being around ~2 μm. In addition, at a higher magnification (10,000 ×), microcracks were clearly seen close

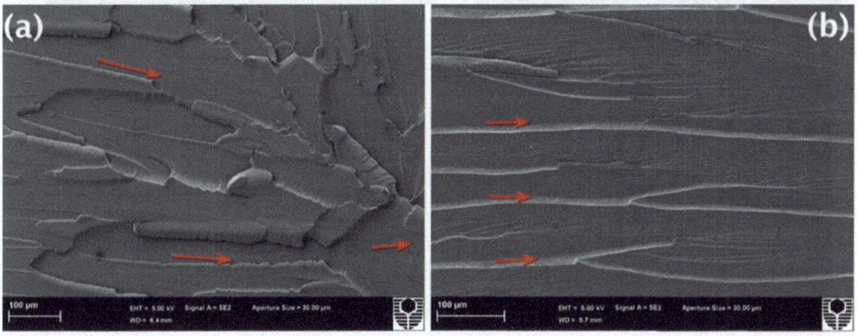

Fig. 5.11 Fracture surfaces of bioepoxy matrices in bionanocomposites at the ESO content of 20 wt% with different curing agents. **a** IPDA and **b** MTHPA. Arrows indicate crack propagation on fracture surfaces

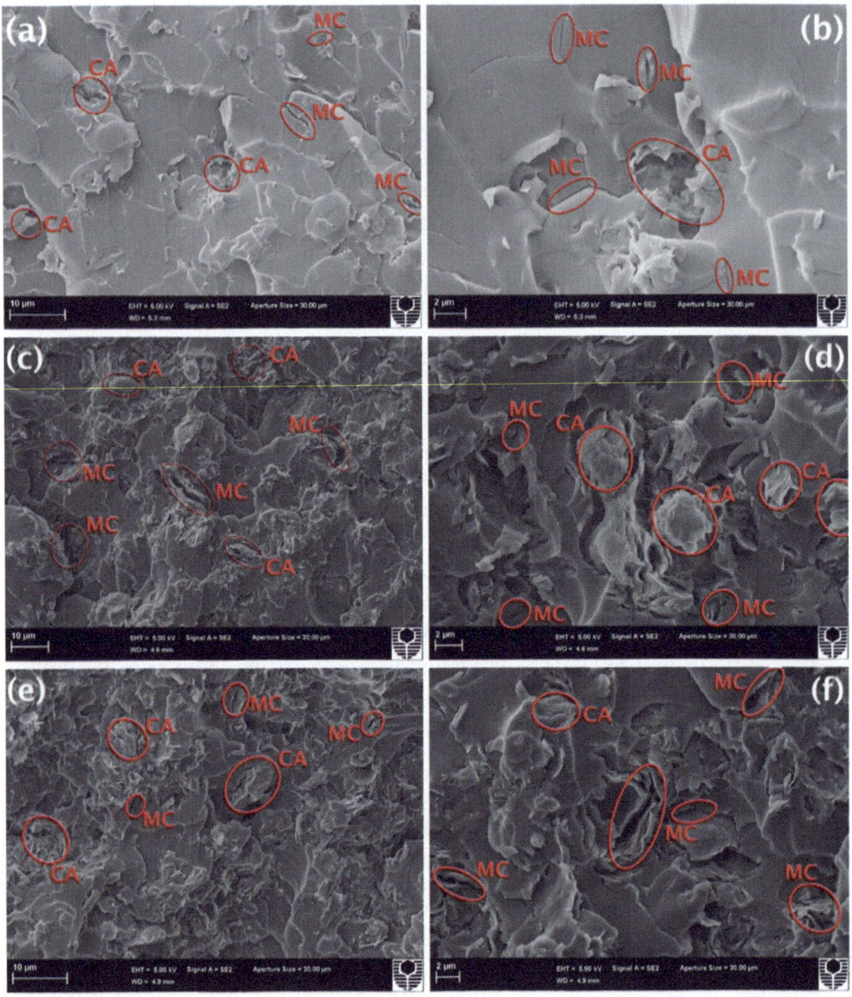

Fig. 5.12 SEM micrographs of fracture surfaces of bionanocomposites cured by IPDA at different clay contents with various magnifications. **a** 1 wt%, at 1,000 × and **b** at 10,000 × ; **c** 5 wt%, at 1,000 × and **d** at 10,000 × , **e**8 wt%, at 1,000 × and **f** at 10,000 × . The letters CA and MC are referred to as clay aggregates and microcracks, respectively [5]

to clay aggregates and localised matrices (Fig. 5.12b). On the other hand, similar fracture surface characteristics were demonstrated at higher clay contents (i.e. 5 wt% and 8 wt%), as depicted in Fig. 5.12c–f. The addition of 5 and 8 wt% Cloisite 15 clays showed poorly dispersed clay particles with a clear sign of some aggregated clay particles in bioepoxy matrices. The aggregates clays were exhibited along the fracture surfaces, which were relatively small in size of ~2 μm (Fig. 5.12c–f). Additionally, it also confirmed the occurrence of microcracks in the fracture surfaces located between bioepoxy matrices and clay aggregates.

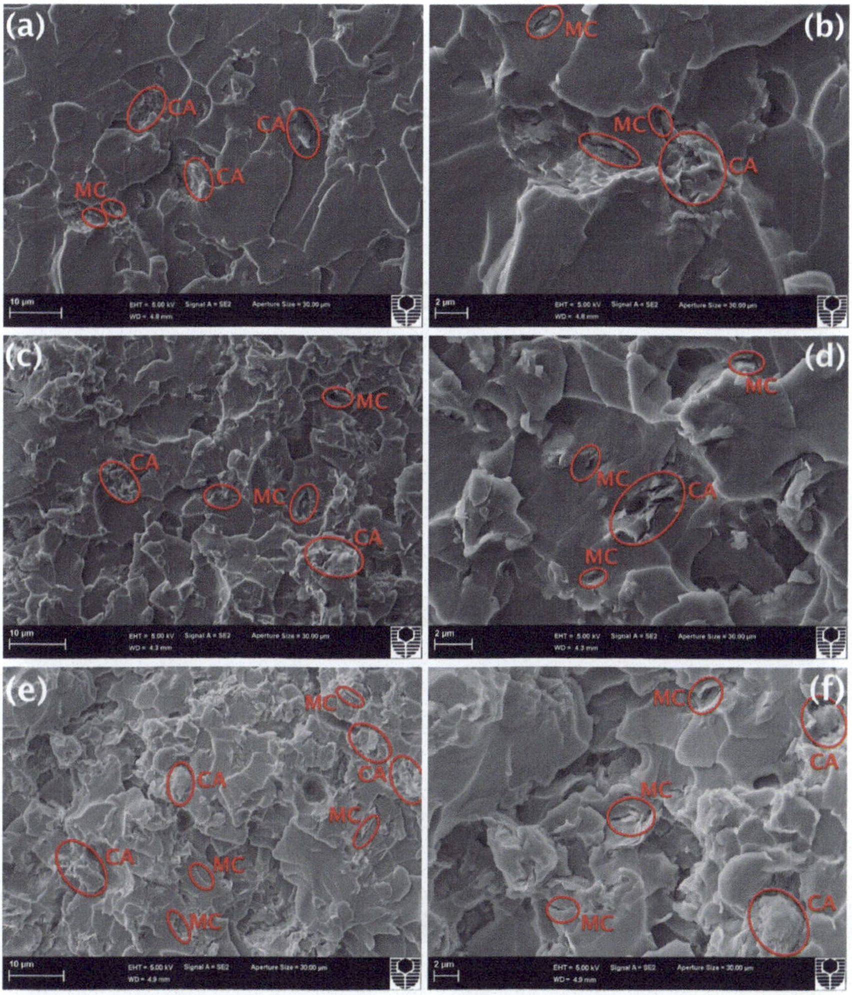

Fig. 5.13 SEM micrographs of fracture surfaces of bionanocomposites cured by MTHPA at different Cloisite 15 clay contents under various magnifications: **a** 1 wt%, at 1,000 × and **b** at 10,000 × ; **c** 5 wt%, at 1,000 × and **d** at 10,000 × ; **e** 8 wt%, at 1,000 × and **f** at 1,000 × . The letters CA and MC are referred to as clay aggregates and microcracks, respectively [5]

On the other hand, fracture surfaces of bionanocomposites cured by MTHPA were also observed at different clay contents, as depicted in Fig. 5.13. A similar fracture surface characteristic of bionanocomposites was demonstrated with rough fracture surface finish, as shown in bionanocomposites cured by IPDA. Clay fillers were well dispersed around fracture surfaces in the presence of some clay aggregates along the surfaces. Moreover, microcracks were observed in the fracture surface of bionanocomposites cured by MTHPA as well, which was likely to appear close to clay aggregates in outer layer sides of clay platelet layers. Nonetheless, as shown

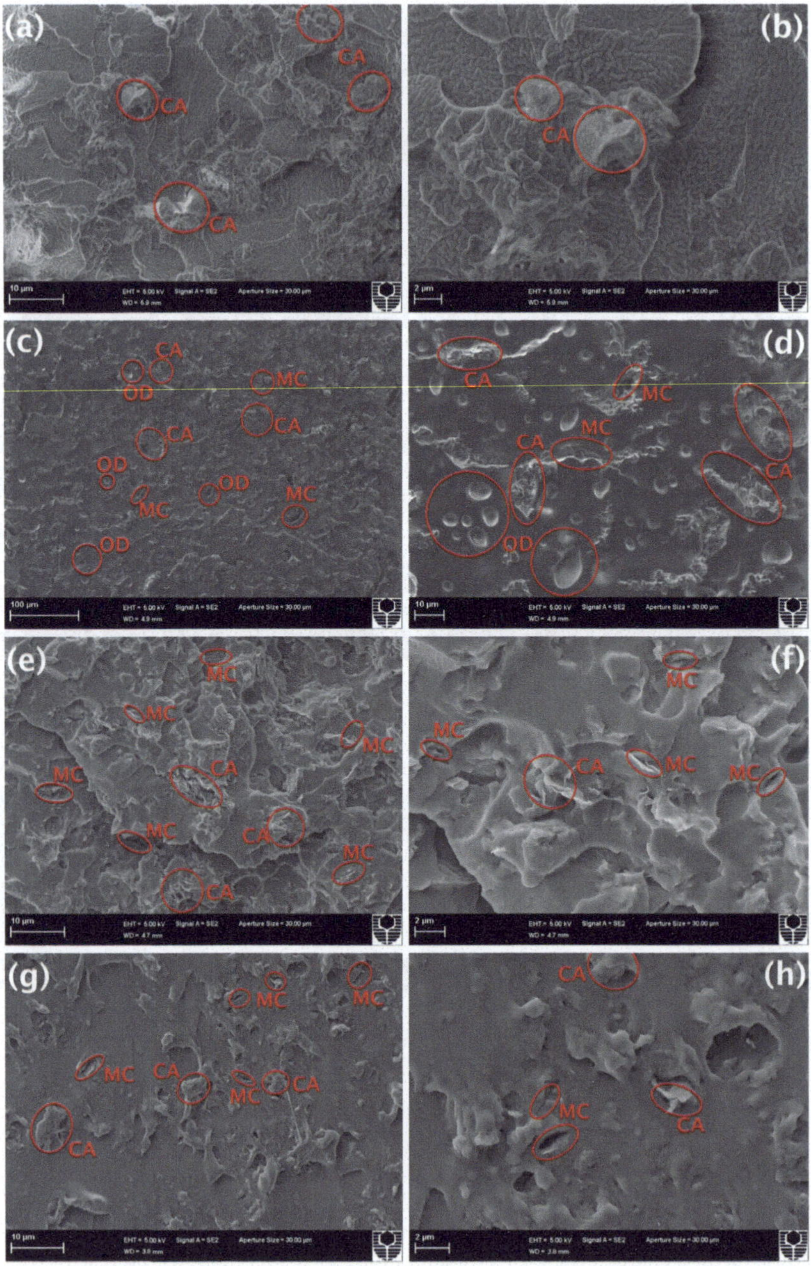

Fig. 5.14 SEM micrographs of fracture surfaces of bionanocomposites reinforced with 5 wt% Cloisite 15 clay with different ESO contents and curing agents under various magnifications: **a** 40 wt% ESO/IPDA, at 1,000 × and **b** at 10,000 × ; **c** 60 wt% ESO/IPDA, at 500 × and **d** at 1,000 × ; **e** 40 wt% ESO/MTHPA, at 1,000 × and **f** at 10,000 × ; **g** 60 wt% ESO/MTHPA, at 1,000 × and **h** at 10,000 × . The letters CA, MC and OD are referred to as clay aggregates, microcracks and oil droplets [5]

in Fig. 5.13c and d, bioepoxy matrices were diffused into clay platelet layers to generate strong interfacial bonding, which was evident from a limited sign of crack propagation around clay aggregates. The initial microcracks near the outer layers of clay platelets could be indicative of stress concentration taking place around clay platelet layers. This phenomenon may be affected by weak interfacial bonding between clay fillers and matrices or in clay platelet layers. However, the size of clay aggregates has noticeable effect on the improvement of mechanical properties [18]. The smallest size of clay particles dispersed in matrices could be correlated to the increase in aspect ratio of fillers. Furthermore, high aspect ratios would enhance specific surface areas of clay fillers in matrices, as well as further improve interfacial bonding between fillers and matrices. As a result, better clay dispersion and interfacial bonding between bioepoxy matrices and clay fillers in bionanocomposites systems would influence resulting mechanical properties, which would be discussed in the forthcoming chapter.

With respect to the effect of ESO content on fracture surfaces, SEM micrographs of bionanocomposites with different ESO contents and curing agents when incorporated with 5 wt% Cloisite 15 clays were depicted in Fig. 5.14. As for bionanocomposites cured by IPDA (Fig. 5.14a–d), increasing the ESO content up to 40 and 60 wt% significantly affected the surface morphology, which demonstrated less roughness and rubber-like material behaviour as compared with those at 20 wt% ESO (Fig. 5.12c–d). Similarly, bionanocomposites cured by MTHPA at the ESO content of 60 wt% possessed smooth surfaces finish with a small size of textured surfaces for embedded clay fillers within the matrices (Fig. 5.12g–h). This typical feature indicated that the addition of 60 wt% ESO could alter material properties of bionanocomposites to be more ductile materials. In contrast, rougher surface finish as a typical sign of brittle materials was clearly identified in fracture surfaces of bionanocomposites cured by MTHPA with 40 wt% ESO, which was similar to those with 20 wt% ESO. As such, ESO content and curing agent type can significantly influence final material characteristics.

On the other hand, bionanocomposites cured by IPDA with 40 wt% ESO (Fig. 5.14a) show good compatibility between bioepoxy blends and the curing agent, leading to homogenous bioepoxy blends. In contrary, the addition of 60 wt% ESO (Fig. 5.14c–d) yielded the fracture surfaces with the second phase of unreacted matrices. The smooth surfaces of bioepoxy matrices and spherical domains of unreacted ESO were clearly shown throughout fracture surfaces. The size of soybean oil droplets was detected to be approximately 2–10 μm in diameter. Such a phenomenon might be influenced by the reactivity of conventional epoxy and ESO with IPDA, which was previously confirmed by FTIR analysis results (Fig. 5.4a). In general, ESO is less reactive than DGEBA because the reactivity from the epoxy ring of ESO is lower compared with those in DGEBA due to the steric hindrance, which is in good accordance with previous studies [6, 19, 20]. The initial epoxy/ESO blends contained the amounts of unreacted soybean oil at higher ESO contents in the DGEBA/IPDA reaction because the rate reaction of ESO/IPDA was lower than DGEBA/IPDA even though some ESO molecules might react with epoxy resin to form chemical bonds at epoxy side groups. In contrast, typical bionanocomposites cured by

MTHPA did not exhibit unreacted ESO phases, as depicted in Fig. 5.13e–h. This finding implied that DGEBA/ESO/MTHPA systems have better compatibility than DGEBA/ESO/IPDA counterparts [21]. The homogenous blending of DGEBA/ESO could react with MTHPA via the polyesterification reaction. In addition, the inclusion of 5 wt% Cloisite 15 clays in bioepoxy matrices with 40 and 60 wt% ESO (Fig. 5.14) also demonstrated non-dispersed clay fillers within polymer matrices in the clay size less than 2 μm on fracture surface areas. Clay aggregates were observed on fracture surfaces to be embedded within the matrices, which was ascribed to crack propagation resulting from non-fully dispersed clay particles in matrices.

5.5 Summary

In this chapter, bionanocomposites were synthesised based on the final preferred combination of factors determined from Taguchi DoE and Pareto analysis of variance (ANOVA). Organomodified clay fillers, namely Cloisite 15 clays were selected based on the optimal material formulation in Taguchi DoE. Meanwhile, different clay contents, ESO contents and curing agent types were used to investigate their effects on morphological structures and fracture surfaces of bionanocomposites. In addition, pre-mixing processing parameters were selected based on Pareto ANOVA. XRD and TEM analyses were employed to investigate clay dispersion statuses within bioepoxy matrices. However, SEM analysis was used to observe fracture surfaces of tensile testing specimens. Moreover, FTIR analysis was employed to examine chemical formulations of resulting bionanocomposites in relation to the effect of clay content, ESO content and curing agent type. Based on the morphological analysis, ESO content influenced the wettability of clay fillers within matrices. The viscosity of bioepoxy blends could affect the molecular mobility to delaminate clay platelets and diffuse them into clay interlayer areas. Based on XRD and TEM analyses, the combinations of partial intercalation/exfoliation, as well as aggregation of clay fillers were observed in bionanocomposites at different clay contents, ESO contents and curing agent types. Nonetheless, the wettability of clay fillers was improved with increasing the ESO content up to 40 wt% when MTHPA was used. On the other hand, the fracture surfaces of bionanocomposites presented a rough surface finish with a sign of brittle materials due to additional clay fillers. Increasing the ESO content could make fracture surfaces become less brittle and more flexible. Interestingly, the functionalisation of curing agents presented different fracture surface finishes. The additional 60 wt% ESO in bionanocomposites cured by IPDA produced two phases of matrices on the surfaces, which corresponded to unreacted soybean oil droplets and cured epoxy/ESO blends. Whereas, those cured by MTHPA curing agent did not yield phase separation. The existence of oil droplets indicated that some unreacted soybean oil appeared in bioepoxy matrices, as confirmed from FTIR spectra, resulting in an increasing and shifting peak absorbance at 1740 cm^{-1} corresponding to stretching $C = O$ of ester in ESO. In contrast, homogenous bioepoxy blends were shown in bionanocomposites when cured by MTHPA with a clear sign of better compatibility between ESO and MTHPA.

References

1. Wang Z, Zhang X, Wang R, Kang H, Qiao B, Ma J, Zhang L, Wang H (2012) Synthesis and characterisation of novel soybean-oil-based elastomers with favorable processability and tunable properties. Macromolecules 45(22): 9010–9019.
2. Salam H (2017) Synthesis and characterisation of bio-based epoxy/clay nanocomposites, PhD thesis, Curtin University, Perth, Australia.
3. Nikolic G, Zlatkovic S, Cakic M, Cakic S, Lacnjevac C, Rajic Z (2010) Fast Fourier transform IR characterisation of epoxy GY systems crosslinked with aliphatic and cycloaliphatic EH polyamine adducts. Sensors 10(1): 684–696.
4. Hashemifard SA, Ismail AF, Matsuura T (2011). Effects of montmorillonite nano-clay fillers on PEI mixed matrix membrane for CO_2 removal. Chem Eng J 170(1): 316–325.
5. Salam H, Dong Y (2019) Property evaluation and material characterization of soybean oil modified bioepoxy/clay nanocomposites for environmental sustainability. Mater Today Sustain 5:100012.
6. Sahoo SK, Mohanty S, Nayak SK (2015) Toughened bio-based epoxy blend network modified with transesterified epoxidised soybean oil: Synthesis and characterisation. RSC Adv 5(18):13674–13691.
7. Singha S, Thomas M J (2008) Reduction of permittivity in epoxy nanocomposites at low nanofiller loadings. Paper presented at the 2008 Annual Report Conference on Electrical Insulation and Dielectric Phenomena, Quebec, Canada, 2008.
8. Fernandes EM, Pires RA, Mano J F, Reis RL (2013) Bionanocomposites from lignocellulosic resources: Properties, applications and future trends for their use in the biomedical field. Prog Polym Sci 38(10–11): 1415–1441.
9. Wade LG (2013) Organic Chemistry. Pearson: Boston.
10. Lan T, Pinnavaia TJ (1994) Clay-reinforced epoxy nanocomposites. Chem Mater 6(12): 2216–2219.
11. Becker O, Cheng Y-B, Varley RJ, Simon GP (2003) Layered silicate nanocomposites based on various high-functionality epoxy resins: The influence of cure temperature on morphology, mechanical properties, and free volume. Macromolecules 36(5):1616–1625.
12. Miyagawa H, Mohanty AK, Misra M, Drzal LT (2004) Thermo-physical and impact properties of epoxy containing epoxidised linseed oil, 2. Macromol Mater Eng 289(7): 636–641.
13. Espinoza-Perez JD, Nerenz BA, Haagenson DM, Chen Z, Ulven CA, Wiesenborn DP (2011). Comparison of curing agents for epoxidised vegetable oils applied to composites. Polym Compos 32(11) 1806–1816.
14. Gu H, Guo J, Wei H, Yan X, Ding D, Zhang X, He Q, Tadakamalla S, Wang X, Ho TC, Guo Z (2015) Transparent anhydride-cured epoxy nanocomposites reinforced with polyaniline stabilised nanosilica. J Mater Chem C 3(31): 8152–8165.
15. Wang R, Schuman T, Vuppalapati RR, Chandrashekhara K (2014) Fabrication of bio-based epoxy-clay nanocomposites. Green Chem 16(4):1871–1882.
16. Bordes P, Pollet E, Avérous L (2009) Nano-biocomposites: Biodegradable polyester/nanoclay systems. Prog Polym Sci 34(2):125–155.
17. Pesetskii SS, Bogdanovich SP, Myshkin NK (2013) Tribological behavior of polymer nanocomposites produced by dispersion of nanofillers in molten thermoplastics. In: Friedrich K, Schlarb AK (eds) Tribology of Polymeric Nanocomposites. Butterworth-Heinemann: Oxford, 119–162.
18. Ahmed S, Jones FR (1990) A review of particulate reinforcement theories for polymer composites. J Mater Sci 25(12): 4933–4942.

19. Frischinger I, Dirlikov S (1993) Toughening of epoxy resins by epoxidised soybean oil. In: Keith R, Anthony, JK (Eds) Toughened Plastics I. American Chemical Society: Washington, 51–489.
20. Altuna FI, Espósito LH, Ruseckaite RA, Stefani PM (2011) Thermal and mechanical properties of anhydride-cured epoxy resins with different contents of biobased epoxidised soybean oil. J Appl Polym Sci 20(2): 789–798.
21. Altuna FI, Espósito L, Ruseckaite RA, Stefani PM (2010) Syntactic foams from copolymers based on epoxidised soybean oil. Compos Part A: Appl Sci Manuf 41(9): 1238–1244.

Chapter 6
Properties of Optimal Material Formulation of Bioepoxy/Clay Nanocomposites

6.1 Mechanical Properties

Mechanical properties of bioepoxy/clay nanocomposites such as tensile and impact properties were summarised with respect to the effect of clay filler content, ESO content, as well as curing agent type, as shown in Figs. 6.1 and 6.2. Tensile strength and tensile modulus of bionanocomposites were considerably improved with the inclusion of clay fillers despite the retention of impact strength. As shown in Fig. 6.1a, tensile strength of bionanocomposites at the ESO content of 20 wt% cured by IPDA and MTHPA increased gradually with the inclusion of Cloisite 15 clays up to 5 wt%. In particular, at the clay content of 5 wt%, tensile strength of bionancomposites reached the maximum strength values, which increased up to approximately 16.26 and 77.61% when compared with those of corresponding neat bioepoxies cured by IPDA and MTHPA, respectively. In contrast, with the further clay inclusion (8 wt% Cloisite 15 clays), the tensile strengths of bionanocomposites exhibited a declining trend. Nonetheless, at the Cloisite 15 clay content of 8 wt%, the tensile strength of bionanocomposites cured by IPDA presented a comparable strength value, which was only 4% lower than that of neat biepoxy. Meanwhile, as for bionanocomposites cured by MTHPA, the incorporation of 8 wt% Cloisite 15 clays demonstrated a reduction in tensile strength to approximately 17.97 MPa, which was equivalent to 18% increase relative to that of neat bioepoxy. This can be attributed to poor filler dispersion in the matrices at higher clay contents, as seen from scanning electron microscopic (SEM) images in Figs. 5.12 and 5.13. The presence of aggregated clays would hinder effective filler-matrix stress transfer in bionanocomposites, which was associated with the weak interfacial bonding between fillers and polymer matrices. In addition, the improvement associated with the tensile strength of bionanocomposites can be correlated with dispersed particle size and interfacial bonding between clay fillers and matrices [1]. As depicted in transmission electron microscopic (TEM) images (Fig. 5.8), non-uniform clay dispersion within bioepoxy matrices was manifested in bionanocomposites cured by IPDA at a high clay content of 8 wt% for

H. Salam and Y. Dong, *Bioepoxy/Clay Nanocomposites*,
https://doi.org/10.1007/978-981-16-7297-2_6

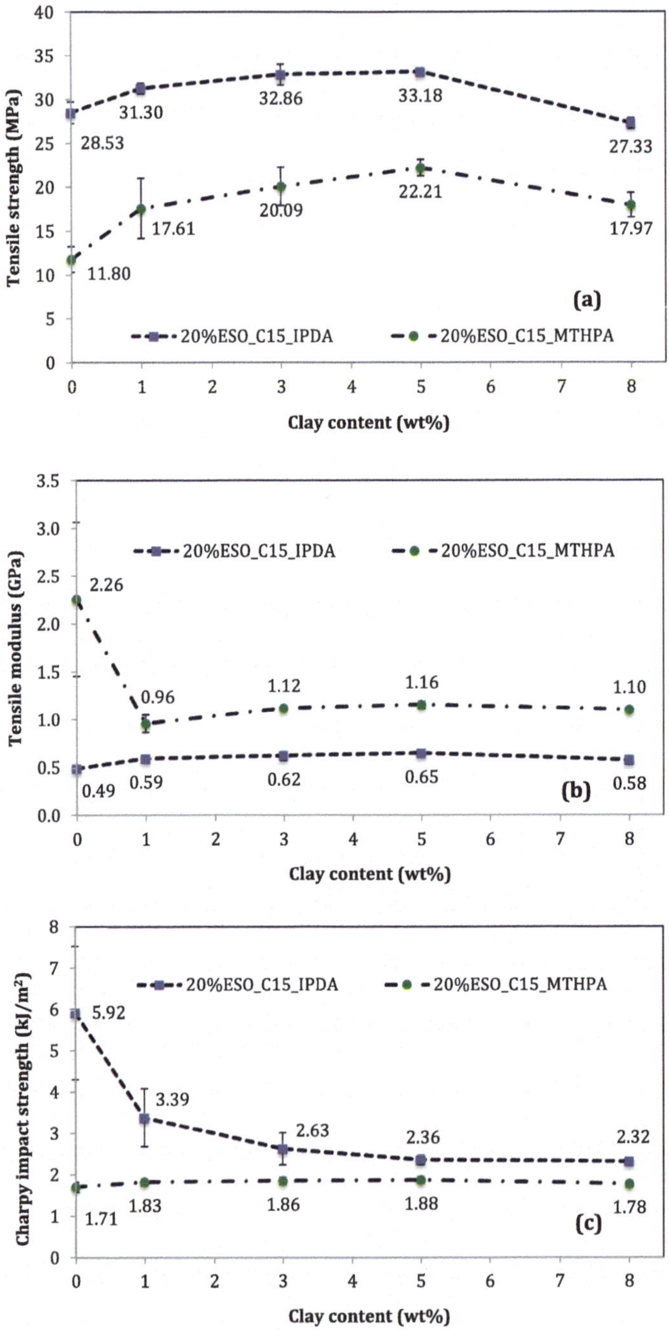

Fig. 6.1 Mechanical properties of bioepoxy/clay nanocomposites at a fixed ESO content of 20 wt% with different clay contents and curing agent types: **a** tensile strength, **b** tensile modulus and **c** Charpy impact strength. The letter C15 is referred to as Cloisite 15 clays

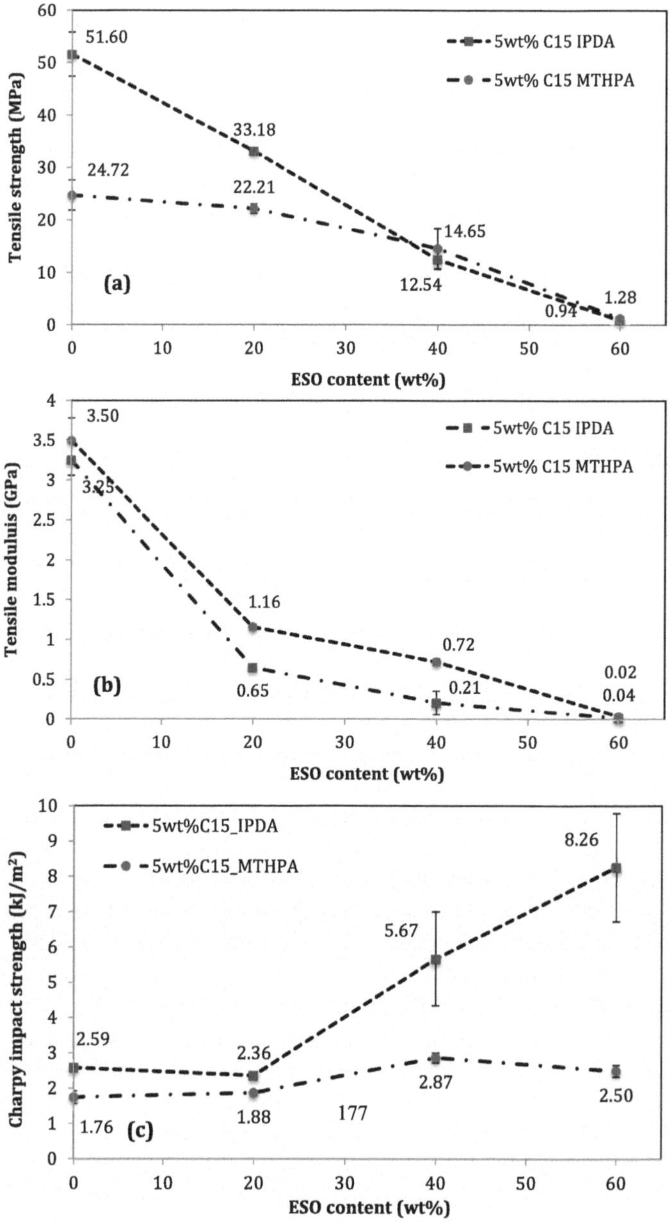

Fig. 6.2 Mechanical properties of bioepoxy/clay nanocomposites reinforced with 5 wt% Cloisite 15 clays with different ESO contents and curing agent types: **a** tensile strength, **b** tensile modulus and **c** impact strength. The letter C15 is referred to as Cloisite 15 clays

Cloisite 15 clays, which induced micro-sized clay aggregates, resulting in the reduction in tensile strength of bionanocomposites. Furthermore, the detrimental tensile strength can be affected by the weak interfacial bonding between fillers and polymer matrices. Conversely, the inclusion of relatively small fillers would further enhance the surface areas of fillers in a composite material system, which could positively influence the effective stress transfer from the matrices to fillers. In this current work, as depicted in fracture surface images in Figs. 5.12, the presence of microcracks at inner and outer areas of clay aggregates was indicative of weak interfacial bonding between matrices and fillers. Furthermore, the declining trend of tensile strength took place. The crosslinking reaction rate of bioepoxy matrices and curing agents could also affect interfacial bonding at inner and outer areas of clay platelet layers. The equilibrium of the crosslinking reaction rate taking place at the clay layers would produce greater interfacial bonding between clay particles and polymer matrices [2, 3]. Based on above-mentioned results, the tensile strength trend for such bionanocomposites can be attributed to the intertwined nexus between particle size and interfacial bonding, as well as between clay particles and bioepoxy matrices, which was in good agreement with previous findings [1, 4].

On the other hand, as shown in Fig. 6.1b, increasing the clay content was also detected to improve the tensile modulus of bionanocomposites at a fixed ESO content of 20 wt%. Bionanocomposites with the inclusion of 1–5 wt% Cloisite 15 clays revealed an increase in tensile modulus from 0.59 to 0.65 GPa, (i.e. improvement from 21 to 33%) when compared with that of neat bioepoxy cured by IPDA. However, when the clay content increased up to 8 wt%, the tensile moduli were slightly reduced to about 0.06 GPa. In contrast, as for bionanocomposites cured by MTHPA, the tensile moduli of bionanocomposites was conversely lower than those of neat bioepoxies. Even though the inclusion of clay fillers from 1 to 5 wt% yielded an enhanced tensile modulus from 0.96 to 1.16 GPa, the tensile modulus of bionanocomposites cured by MTHPA was still well below that of neat bioepoxy. With the further addition of 8 wt% clays, tensile modulus of bionanocomposites cured by MTHPA was slightly reduced to approximately 0.06 GPa. This may suggest that an increase in tensile modulus with the addition of 1–5 wt% Cloisite 15 clays can be associated with the inclusion of rigid fillers with a much higher modulus than that of bioepoxy matrices [1]. Whereas, a tensile modulus reduction for bionanocomposites can be attributed to poor clay dispersion, relatively low aspect ratios and random orientation of clay fillers [5]. According to our TEM images, the availability of clay aggregates especially at higher clay contents can indicate a sign of poor clay dispersion and possibly low filler aspect ratios. In addition, the good agreement of X-ray diffraction analysis and TEM results relating to clay random orientation may suggest that filler orientation can be one of possible factors for the modulus reduction in bionanocomposites.

With respect to impact properties, the effect of clay content on the impact strength of bionanocomposites cured by both IPDA and MTHPA (ESO content: 20 wt%) was presented in Fig. 6.1c. In the former bionanocomposites, impact strength was gradually decreased by increasing the clay content up to 5 wt%, which was followed by its retention at 8 wt%. At the clay content of 5 wt%, the impact strength dropped by 60% as opposed to that of corresponding bioepoxy. In contrast, as for the latter with respect to bionanocomposites cured by MTHPA, impact strength demonstrated a minor improvement with increasing the clay content up to 8 wt%. The maximum impact strength was observed at 5 wt% in clay content with 10% increase as compared to that of neat bioepoxy. This deterioration and minor improvement of impact strength in bionanocomposites cured by IPDA and MTHPA could be possibly ascribed to the addition of clay fillers and the existence of clay aggregates. The addition of rigid clay fillers into bioepoxy matrices affected the reduction in the toughness of nanocomposites, which was likely to reduce the ductility of bioepoxy matrices. In addition, the availability of clay aggregates could also act as the weak zones prone to crack propagation. As observed by SEM analysis in Figs. 5.12 and 5.13, such microcracks were located near surrounding clay aggregates, indicating weak interfacial bonding between clay aggregates and bioepoxy matrices. This phenomenon implied that clay aggregation could hinder applied energy dissipated during the impact tests, resulting in low impact energy due to poor energy absorption.

Additionally, with respect to different curing agents, bionanocomposites at a fixed ESO content of 20 wt% cured by IPDA possessed higher tensile strength and impact strength at different clay contents as compared to those cured by MTHPA. However, tensile modulus of bionanocomposites cured by MTHPA appeared to be much higher when compared to corresponding bionanocomposite counterparts cured by IPDA. MTHPA is an alicyclic anhydride, which is less reactive than amine curing agent such as malonic acid [6]. It requires severe curing conditions like elevated curing temperature to activate zwitterion for a crosslinking reaction. Furthermore, at the ESO content of 20 wt%, bionanocomposites cured by MTHPA is probably less flexible with the nature of brittleness in bionanocomposites as opposed to bionanocomposites cured by IPDA. On the other hand, IPDA is an aliphatic alicyclic polyamine with its reactive amine side groups, which could perform a crosslinking reaction to epoxy networks at ambient temperature. Therefore, with 20 wt% ESO, bionanocomposites cured by IPDA has better tensile strengths, as well as impact strengths than those cured by MTHPA.

On the other hand, with respect to ESO content based on bioepoxy, tensile strength, tensile modulus and impact strength of bionanocomposites reinforced with 5 wt% Cloisite 15 clays as a function of ESO content were shown in Fig. 6.2. In general, tensile strength and tensile modulus of bionanocomposites gradually dropped to lower values with increasing the ESO content in conventional epoxy resin along with the improvement of impact strength. As shown in Fig. 6.2a, bionanocomposites cured by IPDA and MTHPA demonstrated a declining trend for tensile strength with increasing the ESO content. Tensile strength of bionanocomposites cured by IPDA dropped by 98%, when 60 wt% ESO was used to replace conventional epoxy, as compared with conventional nanocomposites without ESO. A similar trend was also demonstrated for bionanocomposites cured by MTHPA. An increase in ESO content up to 60 wt% undermined tensile strength of bionanocomposites by approximately 95% as compared with that of conventional nanocomposites. Additionally, tensile strength of bionanocomposites with 20 wt% ESO cured by IPDA exhibited a 50% higher level of tensile strength than that of bionanocomposites cured by MTHPA. Meanwhile, at the high ESO content levels of 40 and 60 wt%, tensile strength of bionanocomposites cured by IPDA presented low strength values (i.e. 14 and 26% reductions) relative to those cured by MTHPA, respectively. This phenomenon indicated that IPDA was less reactive to ESO as compared to MTHPA, which was in good accordance with previously reported results [6–8]. This is because the curing process of bioepoxy with IPDA was conducted at an ambient temperature with a main reaction between phenyl glycidyl ether groups of epoxy and primary amine. However, the secondary amine reaction is small due to the steric hindrance. In addition, bionanocomposites cured by MTHPA was conducted at the elevated temperature in the curing reaction. This leads to a possible minimal steric hindrance effect of ESO molecules to the further etherification reaction of alkoxide with MTHPA. Moreover, the crosslinking degree of bionanocomposites cured by MTHPA increased at a high ESO content, thus resulting in better tensile strength for bionanocomposites.

On the other hand, bionanocomposites demonstrated a typical reduction in tensile modulus with increasing the ESO content, as shown in Fig. 6.2b. Tensile modulus of bionanocomposites dropped for almost 100% with the additional ESO up to 60 wt%. This result demonstrated that the ESO content significantly affected the modulus values, as also mentioned earlier about its role as the most significant factor to influence manufacturing bionanocomposites. Two factors that possibly influenced tensile modulus are the homogeneity of dispersed material components (i.e. bisphenol A diglycidyl ether (DGEBA), ESO and clay fillers), as well as the alteration of bionanocomposite rigidity due to the addition of fillers or co-monomers [4]. Based on SEM results shown in Fig. 5.14, increasing the ESO content up to 60 wt% induced two phases of cured DGEBA/ESO blends and uncured ESO, which was ascribed to the remaining soybean oil droplets in fractured surfaces. Furthermore, the increase in ESO content also inevitably enhanced the volume fraction of ductile material components into bioepoxy, resulting in a higher level of flexibility. As such, increasing the ESO content could further reduce the rigidity of bionanocomposites accordingly.

In contrast, the impact strengths of bionanocomposites reinforced with 5 wt% Cloisite 15 clays with various ESO contents and curing agent types are shown in Fig. 6.2c. The impact strength of bionanocomposites was improved with increasing the ESO content for both curing agents of IPDA and MTHPA. The impact strength value decreased slightly by approximately 8% with the inclusion of 20 wt% ESO cured by IPDA, as opposed to that of conventional nanocomposites without ESO. Beyond this, the impact strength was found to be improved considerably and reached a maximum value at the ESO content of 60 wt%, which was approximately 219% higher than that of conventional nanocomposites. Additionally, the impact strength of bionanocomposites cured by MTHPA increased gradually with increasing the ESO content along with a maximum value of 2.87 kJ/m with 40 wt% ESO, which was about 63% higher than that of conventional nanocomposites. With the further addition of ESO content up to 60 wt%, the impact strength of bionanocomposites declined slightly to 2.50 kJ/m^2. This improvement of impact strength of bionanocomposites is attributed to the transformation of plasticiser phase of ESO into brittle phase in conventional epoxy, which possibly forms rubber-like polymer blends. Additionally, bionanocomposites cured by MTHPA exhibited lower impact strength values. Such a finding suggested that high stiffness could be obtained with a less rubber-like behaviour as opposed to bionanocomposites cured by IPDA. Apparently, bionanocomposites with the less rubber-like characteristic tended to absorb less dissipated energy accordingly. A further explanation to this in relation to the T_g of bionanocomposites was elaborated in the forthcoming section.

6.2 Thermal Properties

Differential scanning calorimetry (DSC) was carried out to analyse thermal properties such as the T_g of bionanocomposite with respect to the effect of clay content and ESO content, as well as the impact associated with a functional role of different curing agents on bionanocomposites. DSC thermograms of bionanocomposites are shown in Figs. 6.3 and 6.4. Accordingly, Table 6.1 summarised the T_g of bionanocomposites calculated from DSC curves. As for bionanocomposites cured by MTHPA at a fixed ESO content of 20 wt%, additional Cloisite 15 clays improved the T_g of bionanocomposites. In particular, the T_g of bionanocomposites gradually increased with increasing the filler content for Cloisite 15 clays and further reached a maximum level at 8 wt% clay inclusion, which was about 61.99 °C (i.e. 17% higher than that of neat bioepoxy). In contrast, bionanocomposites cured by IPDA presented an apparent retention of T_g values when Cloisite 15 clays were added up to 5 wt%. The T_g of bionanocomposites cured by IPDA with the addition of 8 wt% Cloisite 15 clays was reduced by approximately 8% as opposed to that of neat bioepoxy.

The changes in T_g values of bionanocomposites reinforced with 5 wt% Cloisite 15 clays at different ESO contents were shown in Figs. 6.4. With increasing the ESO content, the thermograms of such bionanocomposites demonstrated the shift of the T_g to a lower temperature level. Nonetheless, as for nanocomposites cured by IPDA (Fig. 6.4a), two glass transition steps of bioepoxy blends appeared with 60 wt% ESO, which could be caused by the availability of two different phases in matrices. This finding suggested the availability of unreacted soybean oil and DGEBA/ESO blend phases according to the SEM observation in Figs. 5.14c–d. The T_g values of bionanocomposites at the clay content of 5 wt% with different ESO contents were summarised in Table 6.1. An increase in the ESO content was found to reduce T_g values of bionanocomposites. The T_g values of conventional nanocomposites (i.e. 0 wt% ESO) cured by IPDA and MTHPA were higher than those of bionanocomposites at the ESO contents in range from 20–60 wt%, which were about 63.69 and 76.60 °C, respectively. In contrary, the T_g of bionanocomposites with 20 wt% ESO was reduced to a minimum value, which was approximately 36 and 23% lower as opposed to corresponding conventional nanocomposites cured by IPDA and MTHPA, respectively. By further increasing the ESO content, the T_g of bionanocomposites slightly increased up to approximately 62–65 °C, despite being still lower than that of conventional nanocomposites. This result implied that the addition of ESO into bioepoxy systems improved the flexibility of bioepoxy blends with a typical rubber-like characteristic. It is well known that ESO is a bio-based

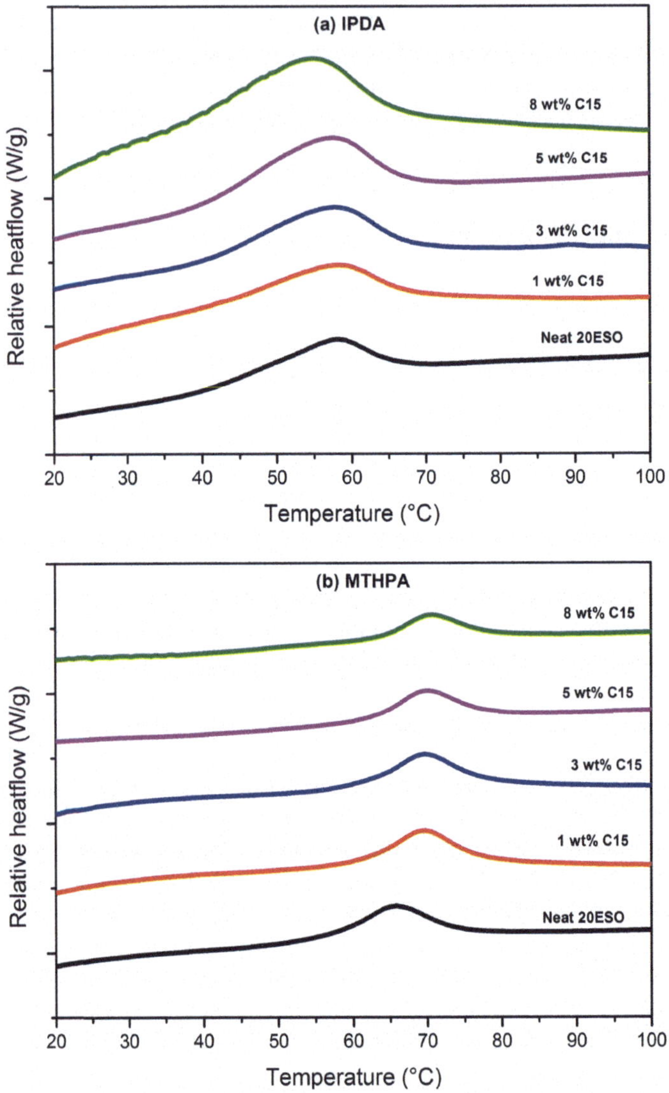

Fig. 6.3 DSC thermograms of bionanocomposites at different clay contents cured by **a** IPDA and **b** MTHPA. The letter C15 is referred to as Cloisite 15 clays [9]

resin with low reactivity, thus leading to a low crosslinking density of cured bioepoxy as compared with that of DGEBA. Nonetheless, the polymerisation reaction of ESO could produce longer polymeric chains between crosslinking points, which became less stable than short-chain counterparts like DGEBA. Furthermore, longer ESO molecular chains easily created the active molecular mobility (DGEBA-ESO) and reduced intermolecular forces in a bioepoxy blend system. Therefore, the substitution

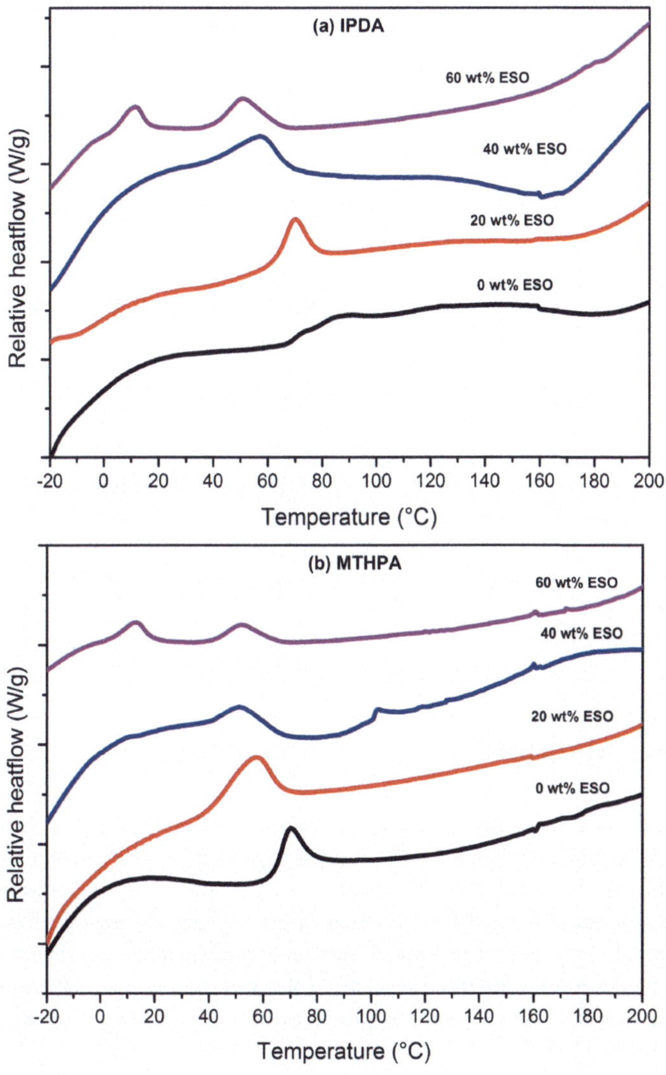

Fig. 6.4 DSC thermograms of bionanocomposites at different ESO contents cured by IPDA and MTHPA [9]

of DGEBA with ESO gave rise to the decrease of T_g owing to an improvement in the chain mobility of blends based on DGEBA/ESO/curing agent. Such a reduction in T_g with increasing the bio-renewable content was in good agreement with those mentioned by Czub [10], suggesting the incorporation of ESO into cross-linked epoxy structures.

Table 6.1 Glass transition temperatures (T_g) of bionanocomposites

ESO content (wt%)	Cloisite 15 clay content (wt%)	Curing agent type	Glass transition temperature (T_g) (°C)
20	0	IPDA	42.05
20	1	IPDA	41.23
20	3	IPDA	41.34
20	5	IPDA	40.69
20	8	IPDA	38.66
20	0	MTHPA	52.79
20	1	MTHPA	53.22
20	3	MTHPA	56.45
20	5	MTHPA	59.10
20	8	MTHPA	61.99
0	5	IPDA	63.69
40	5	IPDA	61.95
60	5	IPDA	58.91
0	5	MTHPA	76.60
40	5	MTHPA	65.64
60	5	MTHPA	63.79

6.3 Water Absorption

Water absorption test results of bionanocomposites were obtained with the effect of clay content, ESO content as well as curing agent types of bionanocomposites, as shown in Figs. 6.5 and 6.6. Figure 6.5 revealed the water absorption rate of bionanocomposites at a fixed ESO content of 20 wt% as a function of immersion time. In general, the water absorption of bionanocomposites increased with an increment of immersion time. Its trend basically followed the Fickian behaviour, which demonstrated an increase in water absorption prior to the saturation with increasing immersion time [11–13].

Bionanocomposites cured by IPDA at a fixed ESO content of 20 wt% exhibited a rapid water absorption rate in a 56-day immersion duration at different clay contents along with further slow water absorption in subsequent immersion time for 180 days according to Fig. 6.5a. In the immersion period over 56 days, the water absorption reached an equilibrium condition in a range from 2 to 2.5%, depending on the clay content. Nonetheless, increasing the clay content revealed an increase in water absorption rate in bioepoxy matrices. Additional 8 wt% Cloisite 15 clays presented a maximum water absorption rate about 2.59%, which was 32% higher than that of neat bioepoxy. In addition, for the immersion time less than ~ 40 days, bionanocomposites at all clay contents revealed a comparable water absorption rate as opposed to that of neat bioepoxy. For instance, in the immersion duration for

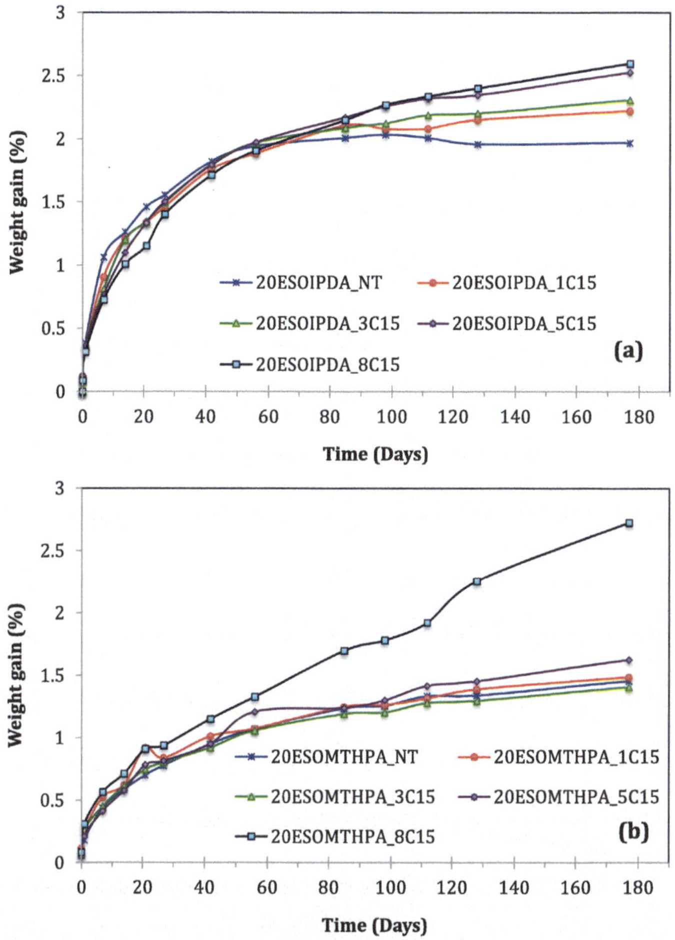

Fig. 6.5 Water absorption of bionanocomposites cured by **a** IPDA and **b** MTHPA with the effect of clay content. The letter C15 is referred to as Cloisite 15 clays

27 days, the water absorption rate of bionancomposites was reduced to a minimal value with increasing the clay content up to 8 wt%, which was about 10% lower than that of neat bioepoxy.

In comparison, water absorption rate of bionanocomposites cured by MTHPA also demonstrated an increasing trend with the increment of immersion time, as shown in Fig. 6.5b. The water absorption rate of bioepoxy and bionanocomposites with the inclusion of 1–5 wt% clays became faster during the immersion time of 80 days. Beyond this, the increase in water absorption rate appeared to slow down with the equilibrium condition obtained at approximately 1.5%. However, bionanocomposites reinforced with 8 wt% clays demonstrated a rapid water absorption rate until the end of immersion time for 180 days and eventually reached the water absorption rate of

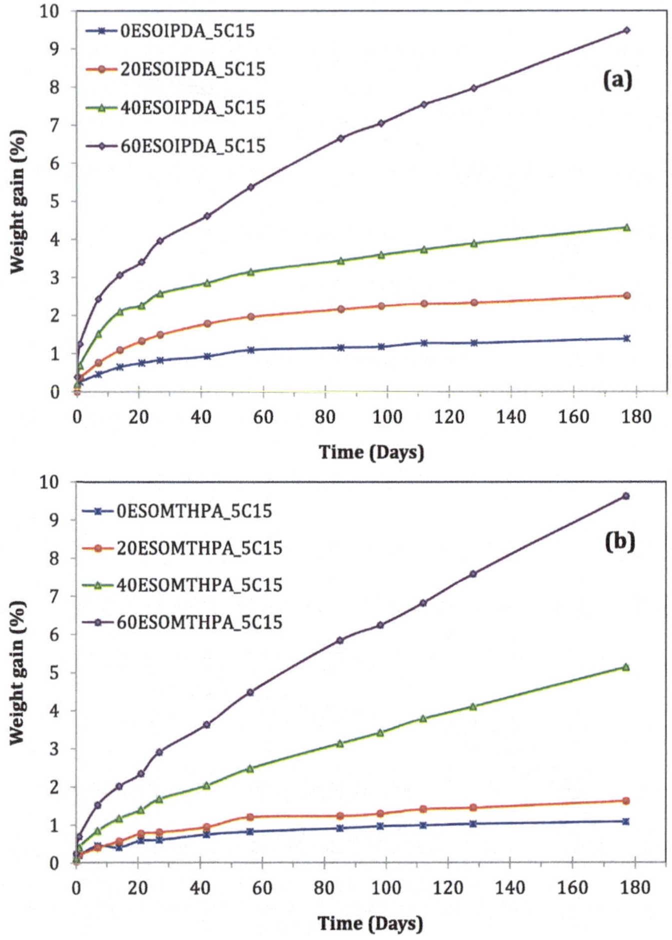

Fig. 6.6 Water absorption of bionanocomposites cured by **a** IPDA and **b** MTHPA with the effect of ESO content. The letter C15 is referred to as Cloisite 15 clays

approximately 2.73%. This was the highest absorption rate as compared with those of bioepoxy and bionanocomposites cured by MTHPA at the clay contents ranging from 1–5 wt%. It is also noted that bionanocomposites reinforced with 3 wt% Cloisite 15 clays yielded a slightly higher absorption rate than that of bioepoxy for less than 40 days in immersion time. However, beyond this period, the water absorption rate of bionanocomposites cured by MTHPA with the inclusion of 3 wt% Cloisite 15 clays was lower than that of neat bioepoxy until the end of immersion time, which appeared to be only 1.41%.

On the other hand, the water absorption rate as a function of ESO content for different curing agents of bionanocomposites were demonstrated in Fig. 6.6. With increasing the ESO content, the water absorption of bionanocomposites increased with different immersion time to achieve an equilibrium state in water content. As for bionanocomposites cured by IPDA depicted in Fig. 6.6a, the water absorption rate of conventional nanocomposites and bionanocomposites with 20 wt% ESO presented a rapid incremental percentage of weight gain in 56 days and reached an equilibrium value in subsequent immersion time, which was found to be approximately 1 and 2%, respectively. In contrast, the substitution of 40 and 60 wt% DGEBA with ESO yielded a rapid increase in water absorption rate without any equilibrium state at the end of immersion time. The water absorption rates of bionanocomposites cured by IPDA at the ESO contents of 40 and 60 wt% were about 4.32 and 9.48%, respectively. The water absorption rate at the ESO content of 60 wt% was about 500% higher than those of conventional nanocomposites. Besides, water absorption rate of bionanocomposites cured by MTHPA as a function of ESO content was illustrated in Fig. 6.6b. A similar trend of increasing water absorption rate was also demonstrated for bionanocomposites cured by MTHPA at the clay content of 5 wt%. In addition, the water uptake levels of conventional nanocomposites and bionanocomposites with 20 wt% ESO cured by MTHPA were enhanced slightly while reached the equilibrium condition after 56 days at the water absorption rates of 1 and 1.5%, respectively. However, the functionalisation of 40 and 60 wt% ESO in bionanocomposites revealed a linearly increasing trend with respect to water absorption rate when increasing the immersion time. The water absorption rates reached 5.15 and 9.63% at the end of immersion time of 180 days, which improved by approximately 370 and 780% as compared with those of conventional nanocomposites.

According to aforementioned results, the behaviour of increasing immersion time for bionanocomposites could be generally identified as an evidence of colour change, as shown in Figs. 6.7 and 6.8. When increasing the clay content, colour change was clearly observed from yellowish clearness before the immersion in water at 0 day to opaque yellow with additional 1–8 wt% Cloisite 15 clays (Fig. 6.7). In addition, when increasing the immersion time for bionanocomposites cured by IPDA and reinforced with 1–8 wt% Cloisite 15 clays, it apparently demonstrated the colour change of specimens from yellowish to white-yellowish colour, which was particularly the case for bionanocomposites reinforced with 3–8 wt% Cloisite15 clays. Conversely, bionanocomposites cured by MTHPA did not apparently demonstrate colour change over immersion periods with the existence of yellowish colour, except for those with the inclusion of 8 wt% Cloisite 15 clays, which resulted in a light change to be more white-yellowish in colour. Similarly, neat bioepoxy specimens with different curing agents did not cause colour change at all. This colour transformation into a solid white or a white-yellowish colour suggested more hydrolytic degradation taking place in bionanocomposites, which was attributed to the diffusion of water molecules. These macrographs have a good agreement with the water absorption test results (Fig. 6.5), indicating increasing the moisture content up to 2% could transform the specimen colour.

It was worth mentioning that colour change also appeared in bionanocomposite specimens as a function of ESO content, illustrated in Fig. 6.8. The colour change of bionanocomposite specimens seemed to be more white-yellowish with increasing the immersion time, especially at bionanocomposites with high ESO content levels.

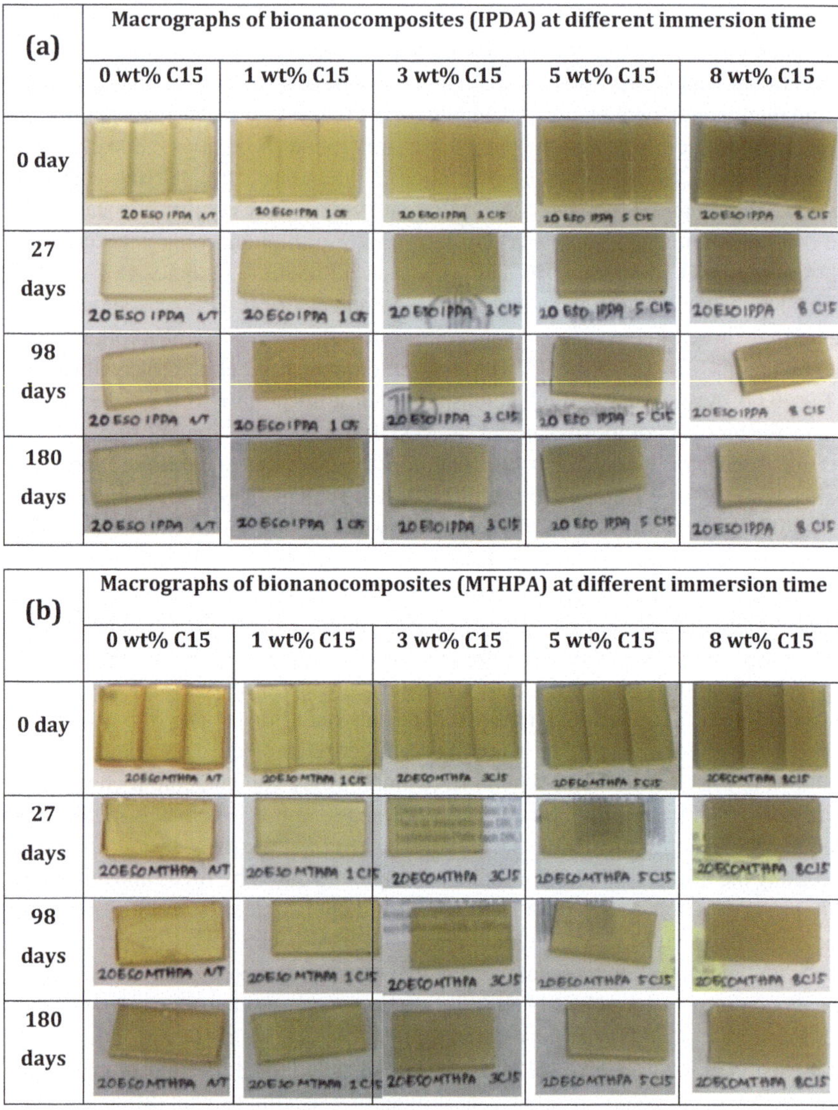

Fig. 6.7 Macroscopic morphology for colour change in bioepoxy blends and bionanocomposites with 20 wt% ESO before and after the immersion in water. Bionanocomposites were cured by **a** IPDA and **b** MTHPA. The letter C15 is referred to as Cloisite 15 clays

After the immersion time for about 27 days, bionanocomposite specimens cured by IPDA demonstrated the colour change to be slightly white-yellowish. Whereas, after the further immersion time, it changed somehow from white-yellowish to solid white colour. This spectrum of colour transformation of specimens from being yellowish-clear to white-yellowish or to solid white colour can be indicative of more water

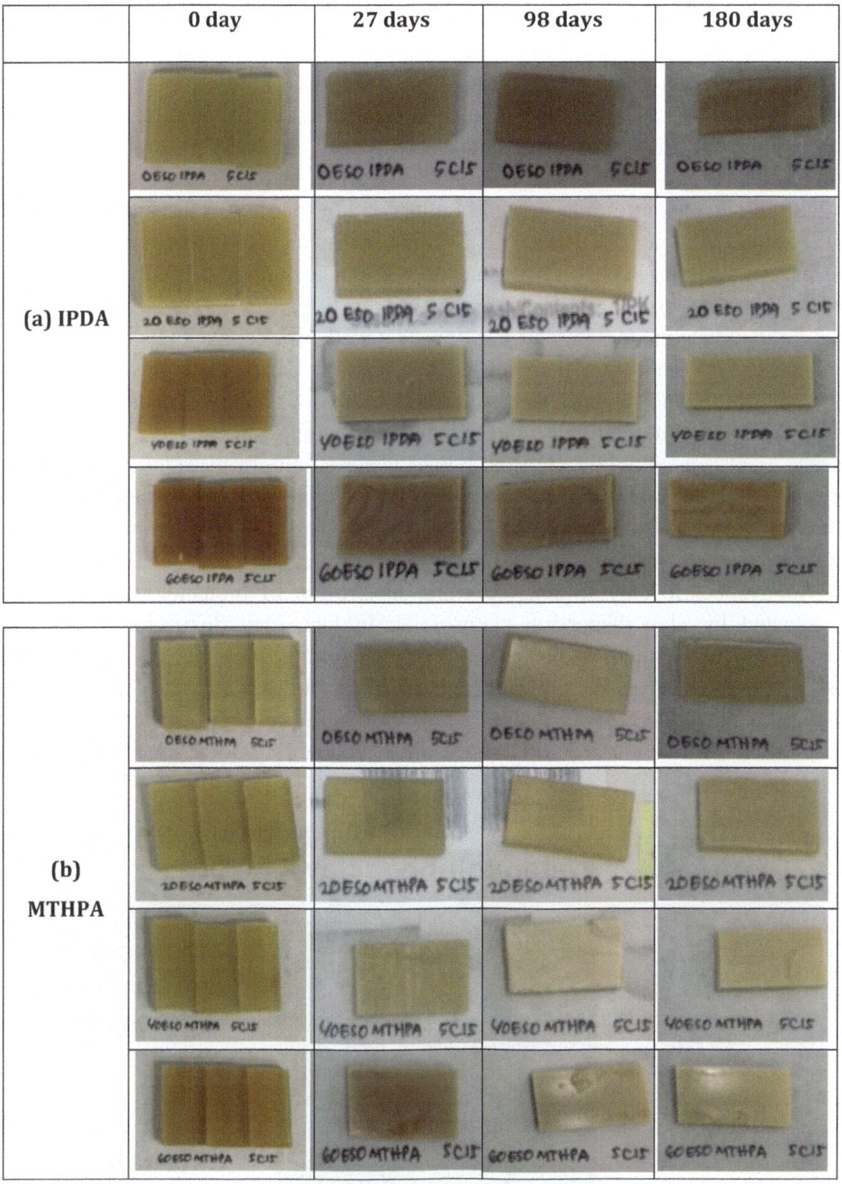

Fig. 6.8 Macroscopic morphology for colour change in bionanocomposites reinforced with 5 wt% Cloisite 15 clays at different ESO contents before and after water immersion. Bionanocomposites were cured by **a** IPDA and **b** MTHPA

diffusion into specimens. Evidently, the white-yellowish colour represented more water diffusion. Therefore, those macrographs were in good agreement with the increasing water absorption results depicted in Fig. 6.6.

The water absorption phenomenon in bionanocomposites could be attributed to the effect of clay fillers, bioepoxy matrices, as well as curing agent. Water diffusion in nanocomposites can be associated with three factors, namely the diffusion of water molecules into matrices, the failure within matrices (i.e. voids or cracks) or the diffusion along the fillers [12, 14]. The diffusion into matrices can be influenced by the polarity of polymeric molecules. Polymers with water active groups (i.e. hydroxyl groups resulting from a crosslinking reaction in epoxy) or a residual hardener could induce the moisture absorption. Water adsorption mechanism in polymer nanocomposites was illustrated in Fig. 6.9. Water molecules can be diffused into nanocomposites and become free water and bound water. Free water stems from water molecules automatically entering into nanocomposite surfaces through the surface cracks or voids. In addition, bound water results from water molecules to form chemical bonding with polar groups of polymer matrices. Nonetheless, in a nanocomposites system, water molecules would also diffuse and interact with a hydrophilic side group of fillers, resulting in intermolecular hydrogen bonding [14].

With respect to clay fillers, the water absorption of bionanocomposites can be affected by clay dispersion status [12]. In fact, clay dispersion status not only influences mechanical properties of bionanocomposites, but also their water absorption. The capability of bionanocomposites to prevent water diffusion is affected by clay structures in resulting bionanocomposites. With better clay wettability in the matrices, it would enhance direct contact surface areas of fillers and

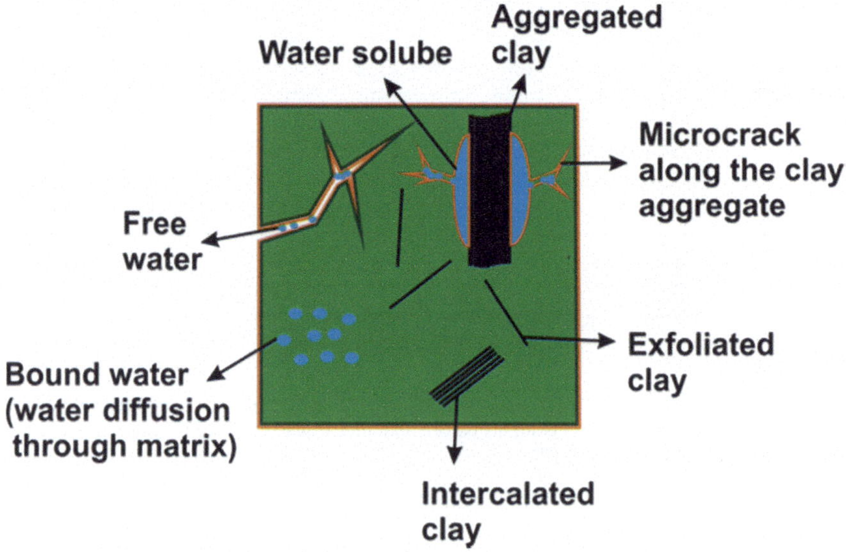

Fig. 6.9 Schematic diagram of water absorption in polymer nanocomposites. Adapted from [14]

matrices. However, poor wettability would reduce such surface contacts, which was likely to produce free volume in the interphase region, potentially filled by water molecules. Exfoliated clay structures can generate a barrier in preventing water molecules in order to diffuse into bionanocomposites [11, 12, 15]. As such, it would produce bionanocomposites with low water absorption rates as compared with neat biopolymer. In contrast, the increase in water absorption rate with additional clays in bionanocomposites can be indicative of the failure surfaces of bionanocomposites or clay aggregation. As previously depicted in Figs. 5.8 and 5.9, clay aggregation has apparently occurred in bionanocomposites. Since clay particles have hydrophilic nature, aggregated clays in bioepoxy matrices may adsorb water during the immersion time. Accordingly, the water absorption rate of bionanocomposites was higher than that of bioepoxy matrices. Additionally, different water absorption rate by increasing the clay content could be a clear sign of different water absorption behaviours of bionanocomposites in terms of the degree of clay aggregation in the matrices. These results were in good accordance with our morphological structure results of bionanocomposites, which manifested an intensification of clay aggregates with increasing the clay content. In addition, the phenomenon for bionanocomposites cured by IPDA in the immersion time less than 40 days can be associated with the diffusion rate of water molecules into bionanocomposites surfaces. This plausible reason might be correlated to the transport rate of water molecules and clay dispersion status on the surfaces of bionanocomposites, which arises from better clay dispersion with exfoliated clay structures to alter the diffusion of water molecules into bionanocomposites. In addition, in the further immersion time, neat bioepoxy reached the saturation level of water absorption while water molecules in bionanocomposites still diffused into bionanocomposites and increased moisture absorption due to the availability of clay aggregates. Furthermore, at the end of immersion time, the water absorption rate of bionanocomposites was higher than that of neat bioepoxy.

On the other hand, with respect to different ESO contents, in addition to the effect of clay structures in bionanocomposites as previously mentioned, increasing water absorption rate by immersion time for bionanocomposites could be attributed to the component compatibility between bioepoxy blends and curing agents such as crosslinking density, unreacted substances (i.e. matrices and curing agent), as well as the diffusion process of water in the matrices [10, 16–19]. The reduction in the crosslinking density of bionanocomposites could result from the plasticisation effect of ESO, which was identified from the decreased T_g results, as listed in Table 6.1. As such, increasing the ESO content reduced crosslinking density, leading to higher water adsorption rate of bionanocomposites relative to conventional nanocomposites. The other possible reason lies in unreacted substances including curing agents and the matrices from an incomplete curing process. Such a curing process can produce hydroxyl groups (-OH) in bioepoxy matrices. During the curing reaction, epoxy groups of both DGEBA and ESO can react with curing agents to produce crosslinking structures with the formation of hydroxyl groups. Since hydroxyl groups contain polar moistures, they would probably interact with water molecules to generate hydrogen bonding. Furthermore, in relation to a non-fully-cured material with the high availability from being free of hydroxyl groups, it also produces more hydrogen bonds, and may increase water absorption rates. In addition, the unreacted curing agent results from the non-stoichiometric ratio of bioepoxy blends/curing agent or incomplete crosslinking reaction. The groups of curing agent molecules would hydrolyse faster than epoxy groups. The final reason is associated with natural water diffusion into bionanocomposites. This can be affected by the availability of cracks, resulting from applied load during the sample preparation or voids/holes originating from unreacted substance such as oil droplets.

The effect of clay content on Shore D hardness of bionanocomposites with 20 wt% ESO was listed in Table 6.2. Shore D hardness of bionanocomposites slightly improved with increasing the clay content for both IPDA and MTHPA as curing agents. A maximum Shore D hardness was reached with the addition of 5 wt% Cloisite 15 clays, which was followed by a decrease in hardness values with a further increase in clay content. Shore D hardness of bionanocompoistes, reinforced with 5 wt% Cloisite 15 clays at the ESO content of 20 wt% and cured by IPDA and MTHPA, increased by 7 and 3%, respectively as opposed to that of neat bioepoxy. On the other hand, Shore D hardness of bionanocomposites at different ESO contents exhibited

Table 6.2 Properties of water absorption rate and shore D hardness of bionanocomposites

ESO (wt%)	Cloisite 15 clay content (wt%)	Curing agent	Water absorption rate at equilibrium state (%)	Shore D Hardness (D)			
				0 month immersion		6 months immersion	
				AVG	STDV	AVG	STDV
20	0	IPDA	1.97	72.70	1.70	69.00	1.26
20	1	IPDA	2.22	75.40	0.84	72.67	2.07
20	3	IPDA	2.31	75.10	0.32	71.33	1.37
20	5	IPDA	2.53	77.80	1.14	71.17	0.75
20	8	IPDA	2.59	75.70	0.67	69.83	1.47
20	0	MTHPA	1.46	83.90	0.74	82.33	0.82
20	1	MTHPA	1.49	86.10	0.57	83.50	0.55
20	3	MTHPA	1.41	86.40	0.52	83.33	1.03
20	5	MTHPA	1.63	86.60	0.52	81.50	1.05
20	8	MTHPA	2.73	84.90	0.57	82.00	0.89
0	5	IPDA	1.56	85.50	0.53	85.33	1.21
20	5	IPDA	2.53	77.80	1.14	71.17	0.75
40	5	IPDA	4.32	41.40	1.43	34.67	2.25
60	5	IPDA	9.48	10.00	1.05	3.17	0.75
0	5	MTHPA	1.09	85.60	0.52	87.67	0.82
20	5	MTHPA	1.63	86.60	0.52	81.50	1.05
40	5	MTHPA	5.15	78.00	0.82	68.33	1.37
60	5	MTHPA	9.63	38.20	1.14	25.00	1.41

Note AVG = average data and STDV = standard deviation

a decreasing trend with increasing the ESO content (Table 6.2). The functionalisation of 60 wt% ESO exhibited Shore D hardness values about 10 and 38.20 D for bionanocomposites cured by IPDA and MTHPA, respectively, which were reduced by 88 and 55% accordingly, as compared with those of conventional nanocomposites. This minimal improvement of hardness values can be attributed to poor clay dispersion in matrices, as previously confirmed by TEM and SEM results with a typical sign of clay aggregation. In addition, the reduction of hardness by increasing the ESO content can be associated with ESO plasticisation effect.

It was also revealed that water absorption tests influenced hardness properties of bionanocomposites, as listed in Table 6.2. It presented a typical reduction of hardness values after the immersion time, as well as Shore D hardness values of bionanocomposites before and after the water immersion. For example, with respect to different clay content and curing agent type, the Shore D hardness of water-absorbed bionanocomposites, with the inclusion of 20 wt% ESO and 5 wt% Cloisite 15 clays cured by IPDA and MTHPA, decreased by 8.5 and 5.9%, respectively, when compared with that of bionanocomposites before water immersion. This value was also regarded as the highest reduction of Shore D hardness as opposed to other clay contents. As for ESO content, a declining trend of Shore D hardness for water-absorbed bionanocomposites was gradually diminished with increasing the ESO content. The decrease in Shore D hardness became higher by increasing the ESO content, in which the highest reduction was detected for such bionanocomposites with the inclusion of 60 wt% ESO and 5 wt% Cloisite 15 clays (i.e. approximately 68 and 35% for those cured by IPDA and MTHPA, respectively). The deterioration of bionanocomposite properties was primarily associated with the diffusion of water molecules resulting in the plasticisation effect in materials. Moreover, this phenomenon also indicated that ESO content made significant effect on hardness properties of bionanocomposites, which revealed the higher deterioration with the increasing of ESO content as compared with increasing the clay content. Remarkable effect of ESO content on water absorption, as well as hardness values suggested that ESO content was the most significant factor to influence mechanical properties of bionanocomposites, in good accordance with previous findings based on Taguchi DoEs in Chap. 4.

6.4 Biodegradability

ESO is a bio-based renewable polymer, derived from agriculture feedstock, which has a potential resource to replace conventional polymers from fossil fuels. Nowadays, the substitution of conventional polymers is not only focused on less use of petroleum-based polymers and the reduction in mechanical properties of bio-based polymers, but also their biodegradability without hazardous effect to human health, as well as the environment. The biodegradability of such bionanocomposites was investigated by calculating weight loss in composting tests. The weight loss was calculated over different burial time from 0 to 180 days. The weight losses of bionanocomposites at different clay contents and ESO contents were depicted in Figs. 6.10 and 6.11.

Figure 6.10 shows the percentage of weight loss for bionanocomposites at different clay contents using Eq. 2.7, which was considered as a function of burial time for bionanocomposites. In general, bioepoxy/clay nanocomposites possessed a slow degradation process with increasing the clay content. The weight loss was reached below 2% for bionanocomposites with 20 wt% ESO cured by IPDA and MTHPA after 180 days in burial time. As for bionanocomposites cured by IPDA shown in Fig. 6.10a, the inclusion of clay fillers demonstrated a decrease in biodegradability as compared with neat bioepoxy. The addition of 1 wt% Cloisite 15 clays exhibited the

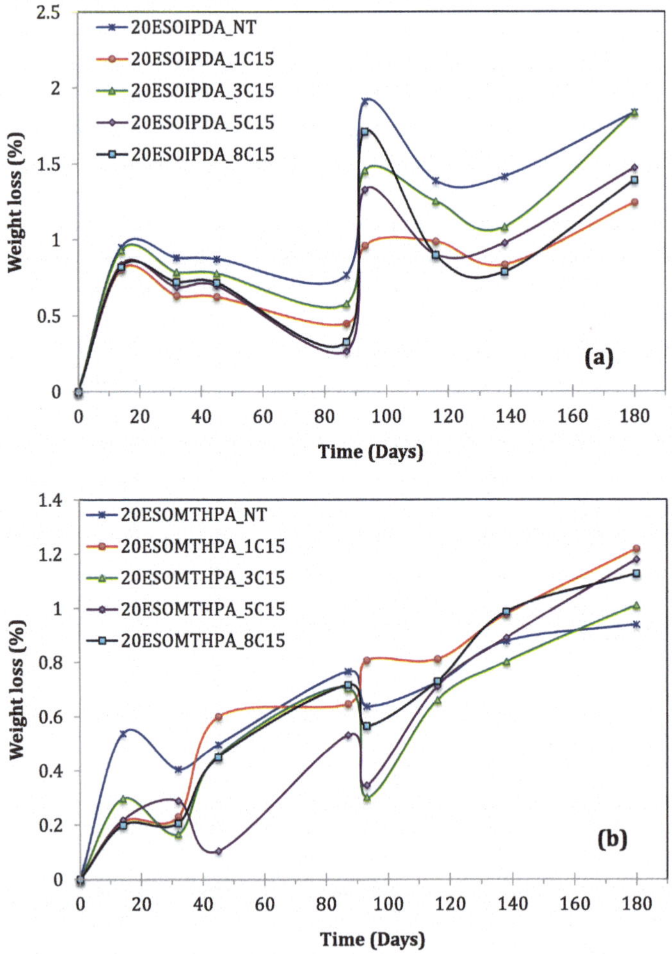

Fig. 6.10 Biodegradability of bionanocomposites with 20 wt% ESO at different clay contents when cured by **a** IPDA and **b** MTHPA. The letter C15 is referred to as Cloisite 15 clays [9]

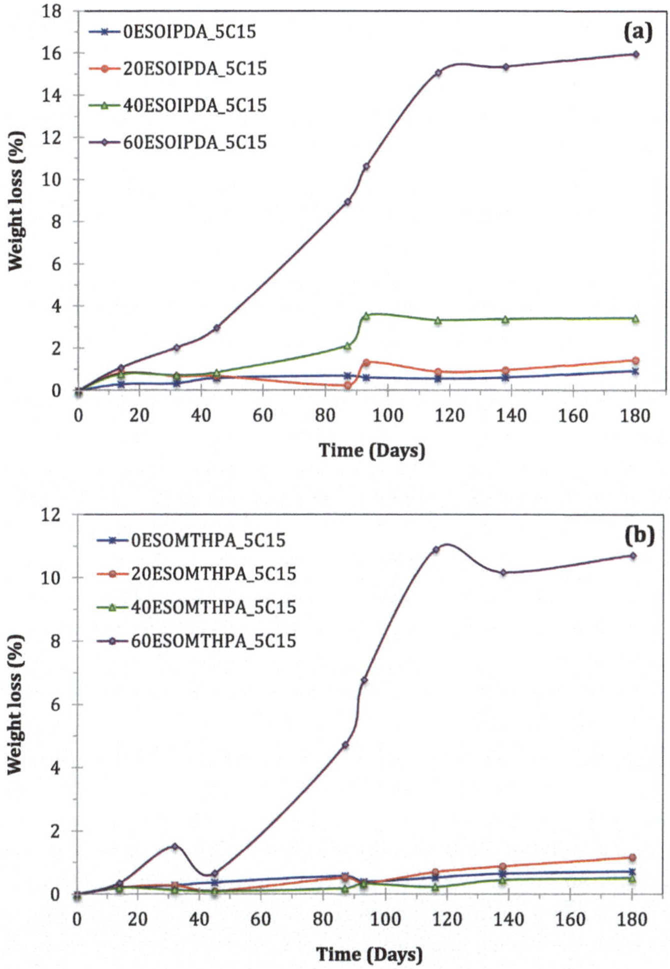

Fig. 6.11 Biodegradability of bionanocomposites reinforced with 5 wt% Cloisite 15 clays at different ESO contents when cured by **a** IPDA and **b** MTHPA. The letter C15 is referred to as Cloisite 15 clays [9]

lowest degradation rate, which was 32% lower than that of neat bioepoxy. A further increase from 5 and 8 wt% clays revealed an increase in weight loss for about 0.23 and 0.15%, respectively, as opposed to that of bionanocomposites reinforced with 1 wt% Cloisite 15 clays. However, this value was still 19 and 24% lower accordingly than that of neat bioepoxy. A maximum weight loss of bionanocomposites was reached by 1.84% with the addition of 3 wt% Cloisite 15 clays when cured by IPDA, which was only 1% higher in comparison with that of neat bioepoxy. In contrast, bionanocomposites cured by MTHPA demonstrated an increasing biodegradability tendency with the addition of clay fillers relative to neat bioepoxy according to

Fig. 6.10b. The addition of 1 wt% Cloisite 15 clays presented the highest weight loss by 1.22% (i.e. 30% higher than that of neat bioepoxy cured by MTHPA). The biodegradability values of bionanocomposites cured by MTHPA were achieved at approximately 26, 20 and 8% at the clay contents of 5, 8 and 3 wt%, respectively. Such levels were still high relative to that of bioepoxy cured by MTHPA.

Moreover, the biodegradability of bionanocomposites reinforced with 5 wt% Cloisite 15 clays was shown in Fig. 6.11. More evidently, the weight loss demonstrated an increasing trend with increasing the burial time for all bionanocomposites at different ESO contents. As for bionanocomposites cured by IPDA given by Fig. 6.11a, the weight loss of conventional nanocomposites appeared to be lower when compared with those of bionanocomposites after 180 days in burial time (i.e. 0.95% compared with 1.5–16%). The substitution of DGEBA with 60 wt% ESO exhibited better biodegradability in 180 days, which was almost 1600% higher than those conventional nanocomposites without ESO at the same clay content of 5 wt%. On the other hand, a similarly nexus was also demonstrated for bionanocomposites cured by MTHPA. A slow biodegradation process was presented for bionanocomposites at the ESO content up to 40 wt%. The weight loss was only increased slightly for about 1% at the end of burial time. However, with a further increase in ESO content up to 60 wt%, the biodegradability of bionanocomposites could be enhanced for about ten folds to 11%, which was 1375% higher as compared with conventional nanocomposites.

Based on above-mentioned results, the incorporation of clay fillers into bioepoxy blends presented different effects, which depended on the functionalisation of curing agents. In particular, the functionalisation of IPDA demonstrated negative impact on the biodegradability of bionanocomposites. Such bionanocomposites showed a relatively low weight loss as compared with neat bioepoxy, resulting from the interaction and the interfacial bonding between clay fillers and matrices. The dispersed silicate clay layers on the surfaces covered bioepoxy molecular chains by obstructing them from the biodegradation motion at the interfaces. However, as for bionanocomposites cured by MTHPA, the biodegradability of bionanocomposite was found to be higher than that of neat bioepoxy even though its value was still lower as opposed to that of bionanocomposites cured by IPDA. In this case, the addition of clay fillers accelerated the degradation process. This behaviour could be attributed to a higher diffusion process of water molecules through hydroxyl groups in bionanocomposites and the improvement of hydrolysis reaction. The materials with higher water solubility can degrade more easily since the microorganism is likely to growth in water [20]. This phenomenon coincides with our previous results based on water absorption tests.

Overall, ESO content had more impact on the biodegradability of bionanocomposites relative to clay content, especially for ESO content being higher than 40 wt%. Increasing the ESO content up to 60 wt% could improve the biodegradability of bionanocomposites by over 10%. The higher biodegradability beyond 40 wt% ESO could be attributed to typical material nature of ESO, which was refined from natural plant oil-based materials. Nonetheless, ESO plasticisation effect can also influence the biodegradability of bionanocomposites. Bionanocomposites at high ESO contents in range from 40 to 60 wt% produced such biomaterials with low crosslinking density due to the existence of longer fatty acid chains among the crosslinking points, which generally appeared in a flexible material at room temperature. Additionally, the weight loss of bionanocomposites cured by IPDA yielded a higher value as compared to those cured by MTHPA. This finding could be attributed to different crosslinking densities of bionanocomposites cured by IPDA and MTHPA. The higher weight loss of bionanocomposite degradation could be associated with the less crosslinking density influenced by the compatibility between bioepoxy matrices and curing agents. This better compatibility can be evidenced from tensile strength and T_g values of bionanocomposites at the high ESO content, as previously illustrated in Fig. 6.2a and Table 6.1, respectively. Regarding bionanocomposites with the inclusion of 60 wt% ESO and 5 wt% Cloisite 15 clays, the tensile strength and T_g of bionanocomposites cured by MTHPA increased by 200 and 8%, respectively, relative to those cured by IPDA. Nonetheless, bionanocomposites with a better crosslinking density would possess lower biodegradability. As such, bionanocomposites cured by MTHPA has lower weight loss than those cured by IPDA.

The degradation process of bionanocomposites in compost media can be influenced by the interaction of bionanocomposites with microorganisms and macroorganisms in soil, as illustrated in Fig. 6.12. Since bionanocomposites is made from non-water-soluble based polymers, the biodegradation of bionanocomposites is regarded as a heterogeneous process. The degradation process starts on the outer surface of bionancomposites because microorganisms are unable to penetrate deeply into the edge of bionanocomposite material as their sizes are too large to penetrate. Furthermore, the biodegradation of polymeric materials usually undergoes a

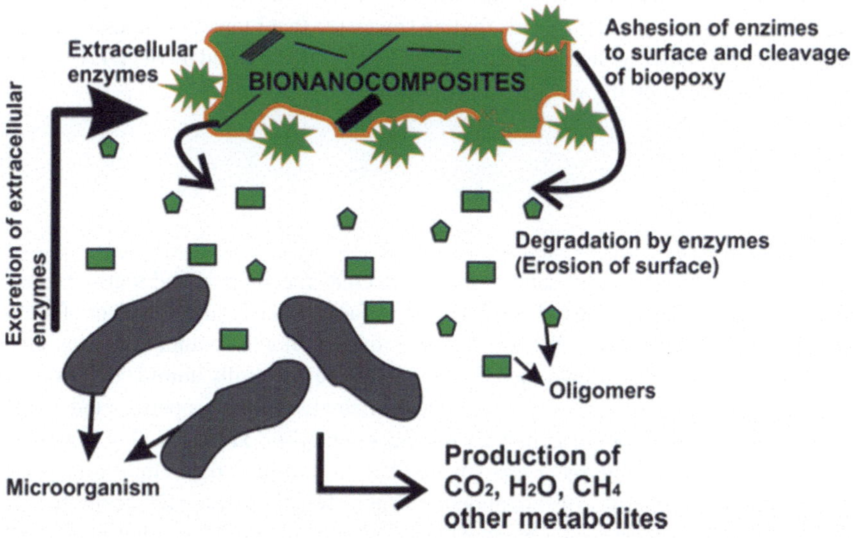

Fig. 6.12 Schematic diagram of biodegradability of bionanocomposites in compost media [21]

surface erosion process. On bionanocomposite surfaces, microorganisms are adhesion and excrete extracelluar enzymes to cleave polymeric chains in the surfaces. This enzyme could break down bioepoxy surfaces through a depolymerisation process. As a result, the molecular weight of bionanocomposites can be reduced, and further can generate water-soluble sides to transport the enzyme into the edge of bionanocomposites for the metabolic reaction. The end product of such metabolism could be anaerobic degradation (i.e. carbon dioxide, water and methane) or aerobic phases. Furthermore, to confirm this theory, SEM analysis was conducted to investigate surface morphology after biodegradability tests. The surface morphology of selected bionanocomposites before and after composting tests was shown in Figs. 6.13 and 6.14.

The surface morphology of bionanocomposites reinforced with 5 wt% Cloisite 15 clays over different periods of 0, 3 and 6 months in soil burial tests were shown in Figs. 6.13 and 6.14. After burial tests, a number of changes and deterioration were observed on the surfaces of bionanocomposites, which was attributed to broken bioepoxy chains during an enzymatic process in soil microorganism. The microorganism in soil attacks bionanocomposite surfaces to alter their surface morphology such as breakage, fragmentation, holes, as well as cranking. Moreover, bionanocomposites surfaces became rough after the burial time. These results had good agreement with the biodegradation mechanism of bionanocomposites, as illustrate in Fig. 6.12. The presence of cracks, holes and fragmentation in bionanocomposites after burial tests was a clear sign of surface erosion. This surface morphology was also reported in biodegradability tests for biopolymers being buried with soil, activated sludge and sweater [5, 22, 23]. Additionally, bionanocomposites cured by IPDA exhibited much clearer alteration after burial tests with the occurrence of breakage and cracking relative to those cured by MTHPA. This finding suggested that bionanocomposites cured by IPDA were relatively easy to degrade as opposed to those cured by MTHPA, which was also confirmed by the weight loss results after burial tests.

6.5 Summary

In this chapter, bionanocomposite properties such as mechanical and thermal properties based on a final preferred combination of factors via Taguchi DoE and Pareto ANOVA were determined. The biodegradability of bionanocomposites was also assessed in order to investigate such bionancomposites in a fully humid condition by means of water immersion, as well as to determine absorption properties and burial characteristic in the compost media for biodegradability. Different clay contents, ESO contents and curing agent types were also used to investigate their effects on aforementioned properties of bionanocomposites. Tensile and impact properties as well as shore D hardness were obtained to look into the material formulation in terms of different clay contents, ESO contents and curing agent types. With respect to clay content, the addition of clay fillers enhanced mechanical and thermal properties of

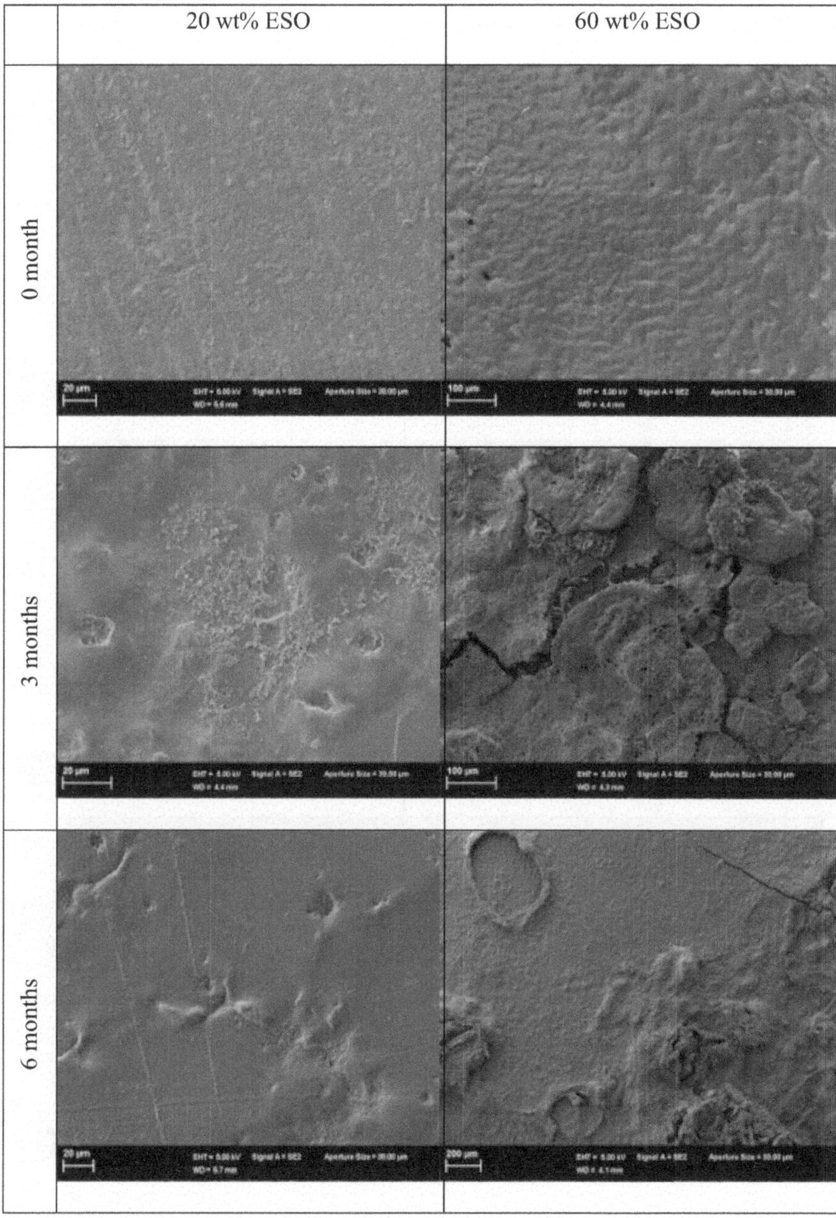

Fig. 6.13 SEM micrographs for surface morphology of bionanocomposites cured by IPDA over different periods of 0, 3 and 6 months in composting tests

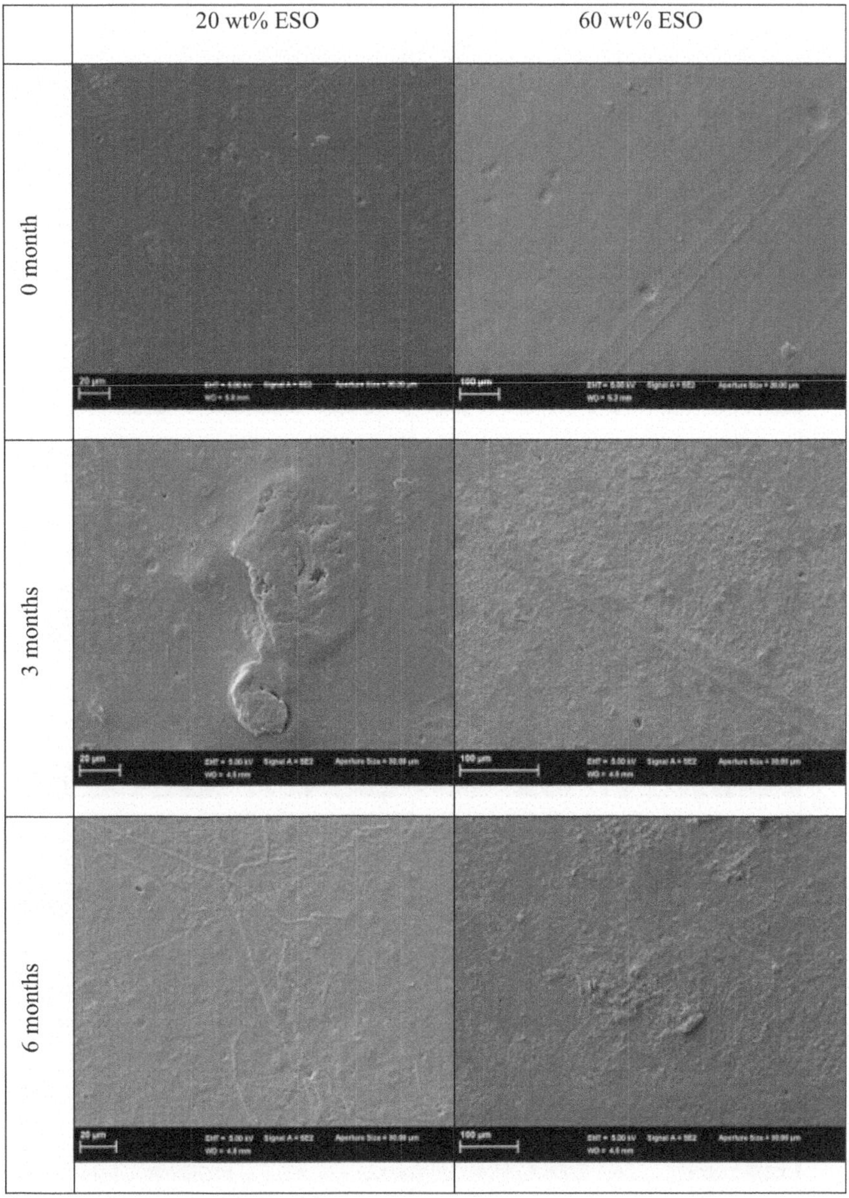

Fig. 6.14 SEM micrographs for surface morphology of bionanocomposites cured by MTHPA over different periods of 0, 3 and 6 months in composting tests

bionanocomposites. The highest tensile strength, tensile modulus and shore D hardness values were revealed by the inclusion of 5 wt% Cloisite 15 clays and 20 wt% ESO for bionanocomposites cured by different curing agents. The further increase in clay content reduced tensile strength and tensile modulus. In addition, the Charpy impact strength of bionanocompoistes cured by IPDA with 20 wt% ESO declined with the addition of clay fillers. Nonetheless, those cured by MTHPA remained comparable to those of neat bioepoxy with increasing the clay content. These phenomena were related to poor clay dispersion and weak interfacial bonding between clays and matrices in bionanocomposites along with the aggregation of clay particles, as confirmed by previous TEM and SEM results. As a result, it can break easily when subjected to stress. T_g values of bionanocomposite materials cured by MTHPA were improved with the inclusion of clays. Meanwhile, those cured by IPDA remained at a lower value as opposed to neat bioepoxy despite its comparable value. Water absorption of bionanocomposites for both curing agents increased with the addition of clay fillers owing to poor clay dispersion in matrices. According to the biodegradability tests, the weight loss was low and almost similar for both bionanocomposites with different curing agent in a range from 1 to 2% in biodegradability rate. This could be caused by the non-water-soluble characteristic of bioepoxy matrices, as well as the dispersion of silicate clay layers on the surfaces of bionanocomposites, thus covering bioepoxy molecular chains from the degradation motion.

With respect to the ESO content, increasing the ESO content from 20 to 60 wt% revealed a declining tendency of tensile properties, hardness and thermal properties of bionanocomposites. Nonetheless, impact strengths appeared to increase with the increasing ESO content. As for bionanocomposites cured by IPDA, tensile strength, tensile modulus, hardness and T_g of bionanocomposites were reduced by 98, 100, 88 and 7%, respectively with 60 wt% ESO. In a similar manner, a relatively high reduction with respect to bionancomposites properties was also found for those cured by MTHPA. The tensile strength, tensile modulus, hardness and T_g of bionanocomposites with 60 wt% ESO decreased by about 95, 100, 55 and 16% in comparison with those of conventional nanocomposites. In addition, the impact strength of bionanocomposites with increasing the ESO content presented a consistent increasing trend and reached a maximum value of 8.26 $kJ \cdot m^{-2}$ when using 60 wt% ESO for those cured by IPDA, as evidenced by the improvement of 220%. Meanwhile, for those cured by MTHPA, the maximum impact strength was achieved by 40 wt% ESO with the enhancement level about 63%. This phenomenon can be attributed to plasticisation effect and steric hindrance of ESO to form crosslinking points between longer aliphatic chains of ESO molecules. Furthermore, increasing the ESO content would produce flexible biomaterials with less brittle nature. In relation to biodegradability, the functionalisation of higher ESO content up to 60 wt% dramatically improved water absorption and accelerated the biodegradation process of bionancomposites. When different curing agents were used (i.e. IPDA and MTHPA), the water absorption rates of bionanocomposites with 60 wt% ESO were improved by about 500 and 780%, respectively, relative to those conventional nanocomposites. Moreover, the biodegradability of bionanocomposites with 60 wt% ESO also demonstrated a remarkable improvement with the increases of 1600 and

1375% as opposed to those of conventional nanocomposites. This phenomenon can be attributed to the availability of hydroxyl groups in bioepoxy matrices, inducing a hydrolysis reaction with water molecules. This is because the microorganism could be developed easily on biomaterial surfaces where surface erosion process took place.

The functionalisation of clay fillers into bionanocomposites systems was found to improve tensile strength, tensile modulus, hardness and T_g of bionanocomposites. It also presented a declining trend of impact strength and biodegradability. In contrast, the increase in ESO content was found to reduce tensile strength, tensile modulus, hardness and T_g of resulting bionanocomposites. The impact strength and biodegradability of bionanocomposites particularly gave rise to a significant improvement. From the viewpoint of curing agent type, bionanocomposites cured by IPDA yielded better impact strength, water absorption and biodegradability as opposed to those cured by MTHPA at all ESO content levels. In contrast, tensile strength demonstrated a better value only at lower ESO contents up to 20 wt%. However, for those cured by MTHPA, more favourable values were obtained for tensile modulus, T_g and hardness at all ESO content levels, as well as tensile strength beyond 40 wt% ESO. Therefore, it could be concluded that overall ESO had better compatibility with MTHPA as opposed to IPDA.

References

1. Fu S-Y, Feng X-Q, Lauke B, Mai Y-W (2008) Effects of particle size, particle/matrix interface adhesion and particle loading on mechanical properties of particulate–polymer composites. Compos Part B: Eng 39(6): 933–961.
2. Kornmann X, Lindberg H, Berglund LA (2001) Synthesis of epoxy–clay nanocomposites: influence of the nature of the clay on structure. Polymer 42(4):1303–1310.
3. Park JH, Jana SC (2003) Mechanism of exfoliation of nanoclay particles in epoxy−clay nanocomposites. Macromolecules 36(8):2758–2768.
4. Ahmed S, Jones FR (1990) A review of particulate reinforcement theories for polymer composites. J Mater Sci 25(12): 4933–4942.
5. Li M-C, Ge X, Cho UR (2013) Mechanical performance, water absorption behavior and biodegradability of poly(methyl methacrylate)-modified starch/SBR biocomposites. Macromol Res 21(7): 793–800.
6. Altuna FI, Espósito LH, Ruseckaite RA, Stefani PM (2011) Thermal and mechanical properties of anhydride-cured epoxy resins with different contents of biobased epoxidised soybean oil. J Appl Polym Sci 120(2): 789–798.
7. Miyagawa H, Mohanty AK, Misra M, Drzal LT (2004) Thermo-physical and impact properties of epoxy containing epoxidised linseed oil, 2. Macromol Mater Eng 289(7): 636–641.
8. Liu ZS, Erhan SZ, Calvert PD (2004) Solid freeform fabrication of epoxidised soybean oil/epoxy composites with di-, tri-, and polyethylene amine curing agents. J Appl Polym Sci 93(1): 356–363.
9. Salam H, Dong Y (2019) Property evaluation and material characterization of soybean oil modified bioepoxy/clay nanocomposites for environmental sustainability. Mater Today Sustain 5:100012.
10. Czub P (2006) Characterisation of an epoxy resin modified with natural oil-based reactive diluents. Macromol Symp 245(1): 533–538.
11. Becker O, Varley RJ, Simon GP (2004) Thermal stability and water uptake of high performance epoxy layered silicate nanocomposites. Eur Polym J 40(1):187–195.

12. Kim J-K, Hu C, Woo RSC, Sham M-L (2005) Moisture barrier characteristics of organoclay–epoxy nanocomposites. Compos Sci Technol 65(5): 805–813.
13. Alamri H, Low IM (2012) Effect of water absorption on the mechanical properties of nano-filler reinforced epoxy nanocomposites. Mater Des 42: 214–222.
14. Azwa ZN, Yousif BF, Manalo AC, Karunasena W (2013) A review on the degradability of polymeric composites based on natural fibres. Mater Des 47: 424–442.
15. Mohan TP, Kuriakose J, Kanny K (2012) Water uptake and mechanical properties of natural rubber–styrene butadine rubber (nr-sr)–Nanoclay composites. J Ind Eng Chem 18(3): 979–985.
16. Perrin-Sarazin F, Sepehr M, Bouaricha S, Denault J (2009) Potential of ball milling to improve clay dispersion in nanocomposites. Polym Eng Sci 49(4): 651–665.
17. El Yagoubi J, Lubineau G, Roger F, Verdu J (2012) A fully coupled diffusion-reaction scheme for moisture sorption–desorption in an anhydride-cured epoxy resin. Polymer 53(24): 5582–5595.
18. Frank K, Childers C, Dutta D, Gidley D, Jackson M, Ward S, Maskell R, Wiggins J (2013) Fluid uptake behavior of multifunctional epoxy blends. Polymer 54(1): 403–410.
19. Toscano A, Pitarresi G, Scafidi M, Di Filippo M, Spadaro G, Alessi S (2016) Water diffusion and swelling stresses in highly crosslinked epoxy matrices. Polym Degrad Stab 133: 255–263.
20. Massardier-Nageotte V, Pestre C, Cruard-Pradet T, Bayard R (2006) Aerobic and anaerobic biodegradability of polymer films and physico-chemical characterisation. Polym Degrad Stab 91(3): 620–627.
21. Souza VGL, Fernando AL (2016) Nanoparticles in food packaging: Biodegradability and potential migration to food—A review. Food Packag Shelf Life 8: 63–70.
22. Ikada E (1999) Electron microscope observation of biodegradation of polymers. J Environ Polym Degrad 7(4): 197–201.
23. Kadam A, Pawar M, Yemul O, Thamke V, Kodam K (2015) Biodegradable biobased epoxy resin from karanja oil. Polymer 72: 82–92.

Chapter 7
Theoretical Modelling of Bioepoxy/Clay Nanocomposites

7.1 Theoretical Models of Polymer/Clay Nanocomposites

Theoretical modelling and numerical simulation techniques of polymer nanocomposites system can be divided into three categories based on the size effect, namely molecular scaled and microscaled methods, as well as mesoscaled and macroscaled methods [1]. Molecular scaled methods are considered as specific modelling approaches for atoms, molecules or clusters of the units. These methods comprise molecular dynamics method, Monte Carlo method and molecular mechanics. However, microscaled methods are used to connect the gap between molecular scaled method and meso/macroscaled methods. Composite constituents are not only affected solely on a molecular scaled level, but also at a microscopic scale to investigate the evolution of structures, bulk flow of materials and the interaction between matrices and fillers. Various methods such as Brownian dynamics, dissipative particle dynamics, lattice Boltzmann, time-dependent Ginsburg-Landau theory and dynamic density functional theories have been well developed to evaluate microscopic structures and the interaction of composite constituents. The last one is meso/macroscaled method, also known as continuum method. Such a method is implemented under the combination condition of both aspects of molecular structures and material nature with the homogenisation at different scaled levels. It concentrates on the deformation of composite materials influenced by externally applied load and their resulting stresses and strains. Macroscaled methods follow essential laws for the continuity of mass, the equilibrium of force, energy and momentum, as well as conservation of energy and entropy [1]. Modelling and numerical simulation have been used to evaluate continuum methods including micromechanical modelling, equivalent-continuum modelling, self-consistent modelling and finite element analysis. In the following sections, we only discussed the application of micromechanical modelling in predicting tensile properties of bioepoxy/clay nanocomposites in terms of tensile modulus and tensile strength based on the effect of individual nanocomposite constituents.

H. Salam and Y. Dong, *Bioepoxy/Clay Nanocomposites*,
https://doi.org/10.1007/978-981-16-7297-2_7

7.2 Modulus of Particulate Polymer Composites

7.2.1 Rule of Mixture (ROM)

ROM model is a basic concept for micromechanical modelling used in polymer particulate composites. The assumptions applied to this model comprise linearly elastic fillers and polymer matrices, the symmetry of fillers including shape, size, as well as aspect ratio and good interfacial bonding between fillers and matrices [1].

Voigt model and Reuss model are the two extreme ROM models as the upper and lower bounds, respectively in theoretical predictions. Voigt model is the combination of overall composite constituents that are influenced by each average modulus and volume fraction of the constituent, which is thus usually called parallel model [2]. In addition, it is also assumed that in the upper bound an isostrain condition exists for both fillers and matrices. As such, the modulus of composites can be written as

$$E_c = E_p\phi_p + E_m(1 - \phi_p) \tag{7.1}$$

In addition, Reuss model (i.e. series model), also called inverse rule of mixtures or lower bound model, is based on the assumption that the stress is uniform in both matrices and fillers [3]. Reuss model can be written in the following form

$$E_c = \frac{E_p E_m}{E_m\phi_p + E_p(1 - \phi_p)} \tag{7.2}$$

and

$$\phi_p = \frac{W_p}{W_p + (\rho_p/\rho_m)(1 - W_p)} \tag{7.3}$$

$$\rho_{blend} = \rho_1\varphi_1 + \rho_2\varphi_2 \tag{7.4}$$

where ϕ_p is the volume fraction of fillers in composites, W_p is the weight fraction of fillers and ρ_m and ρ_p are the densities of matrices and fillers, respectively, which can be obtained from the technical data sheet of the material supplier (Table 2.1). For the density of matrix blends (ρ_{blend}), it is calculated according to Eq. 7.4 that is influenced by the density and volume fraction of each matrix constituent. ρ_1 and ρ_2 are the densities of constituents 1 and 2 while φ_1 and φ_2 are their corresponding volume fractions accordingly. In addition, E_c, E_m and E_p are the moduli of composites, matrices and fillers, respectively. In this study, E_m of bioepoxy blends was determined based on the experiment results obtained from tensile testing of bioepoxy blends at different ESO contents.

On the other hand, the elastic modulus of fillers (E_p) is based on clay dispersion status in bioepoxy matrices. For exfoliated clay structures, E_p is given by 178 GPa, which is the elastic modulus of single montmorillonite (MMT) clay platelet layer [4]. In addition, when intercalated clay structure is formed in matrices, the estimation of

Table 7.1 Effective modulus of intercalated clay particles in bionanocomposites (ESO content: 20 wt%) [23]

Clay filler content	Bionanocomposites with 20 wt% ESO	
(wt%)	d_{002} (nm)	$E_{intercalation}$ (GPa)
1	1.97	85.06
3	1.96	85.59
5	1.96	85.59
8	1.96	85.59

the effective modulus of clay fillers would change, and can be calculated by applying ROM model [4, 5] as follows

$$E_{intercalation} = E_{MMT}\phi_{MMT} + E_{gallery}(1 - \phi_{MMT}) \tag{7.5}$$

$$\begin{aligned} &if\, E_{gallery} \ll E_{MMT} \\ &E_{intercalation} \approx E_{MMT}\phi_{MMT} \approx \left(\frac{d_{MMT}}{d_{002}}\right)E_{MMT} \end{aligned} \tag{7.6}$$

where ϕ_{MMT} and E_{MMT} are the volume fraction in stacks and the modulus of MMT, respectively and $E_{gallery}$ is the gallery modulus. When the gallery modulus is much smaller than MMT modulus ($E_{gallery} < < E_{MMT}$), $E_{gallery}$ would be less significant to the modulus of intercalated clay structures. Accordingly, the modulus of intercalated clay structures can be approximately determined by using Eq. 7.6 in terms of *d*-spacing being calculated from XRD results. Based on XRD and TEM results, the clay structures have been formed with the combination of dominantly intercalated, exfoliated and aggregated clay structures. The interlayer spacing d_{002} of clay fillers was selected demonstrating an interlayer-spacing shift of clay platelet layers. The determination of an effective modulus for intercalated clay fillers at different filler contents and ESO contents are demonstrated in Table 7.1.

7.2.1.1 Modified Rule of Mixture (MROM)

Generally, ROM model is only influenced by moduli and volume fractions of each constituent, in which filler shape, orientation and 3-D direction are ignored. Furthermore, ROM model is modified by introducing the modulus reduction factor (MRF) to estimate the modulus of composites [6]. The modification of ROM model is implemented for the consideration of imperfect filler orientation in the matrices with the inclusion of MRF. Nonetheless, MRF should appear together with the aspect ratio (α) [7]. As such, modified ROM (MROM) model can be written as

$$E_c = E_p\phi_p(MRF) + E_m(1 - \phi_p) \tag{7.7}$$

Two different forms are used for the MRF of flake-like fillers, which is based on Riley's rule [8] and Padawer and Beecher's rule [9], respectively. The MRF equation can be written as follows

For Riley's rule:

$$(MRF) = 1 - \frac{\ln(u+1)}{u} \tag{7.8}$$

For Padawer and Beecher's rule:

$$(MRF) = \frac{\tanh u}{u} \tag{7.9}$$

with

$$u = \frac{1}{\alpha}\sqrt{\frac{\phi_p G_m}{E_p(1-\phi_p)}} \tag{7.10}$$

where α is the aspect ratio and G_m is the shear modulus of matrices. Wu et al. [7] investigated the MRF of rubber/clay nanocomposites by fitting corresponding experimental data. It was found that $MRF = 0.66$ was better fitted to experimental data for the modulus prediction at the filler volume fractions less than 6 vol%.

7.2.1.2 Hirsch Model

Hirsch [10] proposed a model for estimating the modulus of composites with the combination of parallel and series models. The general form of such an equation is the effect of the modulus of matrices, the modulus of fillers, their volume fractions along with an empirical constant x. Hirsch model equation is given by

$$E_c = x\left(E_m(1-\phi_p) + E_p\phi_p\right) + (1-x)\frac{E_p E_m}{E_m\phi_p + E_p\left(1-\phi_p\right)} \tag{7.11}$$

where x is an empirical constant to control the stress transfer between fillers and matrices, which is based on the curve fitting with experimental data ($0 < x < 1$).

7.2.1.3 Halpin–Tsai Model

Halpin–Tsai (H-T) model [11, 12] offers a reasonable prediction for the elastic modulus of unidirectional composites with respect to volume fraction and filler geometry. H-T model can be used for continuous or discontinuous fillers such as fibre-like or flake-like fillers with different alignment directions. The elastic modulus of composites can be written in the following form

$$E_c = E_m \frac{1 + \xi \eta_L \phi_p}{1 - \eta_L \phi_p} \tag{7.12}$$

where E_c, E_m and E_p represent the elastic moduli of composites, matrices and fillers, respectively, and η_L is given by:

$$\eta_L = \frac{(E_p/E_m) - 1}{(E_p/E_m) + \xi} \tag{7.13}$$

ξ is a constant that depends on the geometry and aspect ratios of fillers in composites and α is the aspect ratio. ξ can be expressed in the following forms [4]

$$\xi = 2\alpha = 2\left(\frac{l}{t}\right) \quad for\,longitudinal\,modulus(E_{11}) \tag{7.14}$$

$$\xi = 2 \quad for\,transverse\,modulus(E_{22}) \tag{7.15}$$

where the parameters l and t are the length and thickness/depth of dispersed fillers in composites, respectively. In addition, when the ξ value is very large ($\xi \rightarrow \infty$), H-T model is equal to ROM model (i.e. parallel model), as given in Eq. 7.1. Contrarily, when the ξ value is very small ($\xi \rightarrow 0$), H-T model becomes an inverse model of ROM (i.e. series model), as given in Eq. (7.2). Consequently, H-T model can be used for most prediction situations between lower and upper limits in ROM.

On the other hand, in polymer/clay nanocomposites system, H-T model can be used to predict elastic moduli of nanocomposites influenced by the volume fraction and aspect ratio of clay fillers, as well as the formed structures of clay fillers in matrices [4, 5, 13]. In completely exfoliated clay structures, the prediction of elastic modulus of nanocomposites would be calculated using Eq. 7.12 with E_p/E_m equal to the ratio of the modulus of a single clay platelet layer to the modulus of matrices. In addition, for an intercalated clay structure system, elastic modulus prediction is made according to Eq. 7.12, but E_p/E_m can change and become the ratio of elastic modulus of platelet stacks to that of matrices. The elastic modulus of intercalated clay structures can be calculated based on Eq. 7.6. On the contrary, for randomly oriented clay platelets, the elastic modulus of nanocomposites is determined based on the combination of longitudinal (E_{11}) and transverse (E_{22}) moduli of nanocomposites with the same aspect ratio and volume fraction of clay fillers using the following [13, 14]:

$$E_c = E_m \left[\frac{3}{8}E_{11} + \frac{5}{8}E_{22}\right] \tag{7.16}$$

or

$$E_c = E_m\left[\frac{3(1+\xi\eta_L\phi_p)}{8(1-\eta_L\phi_p)} + \frac{5(1+2\eta_T\phi_p)}{8(1-\eta_T\phi_p)}\right] \tag{7.17}$$

with

$$\eta_T = \frac{(E_p/E_m)-1}{(E_p/E_m)+2} \tag{7.18}$$

$$\xi = \frac{2}{3}\alpha = \left(\frac{2l}{3t}\right) \tag{7.19}$$

Additionally, 2D structure shape of clay platelets can possibly make less significant contribution to the modulus of nanocomposites as compared with those filled by 1D fibre-like counterparts. Hence, MRF for clay platelet fillers is utilised in modified H-T model based on Eqs. 7.12 and 7.17, which are applied below accordingly

$$E_c = E_m\frac{1+\xi(MRF)\eta_L\phi_p}{1-\eta_L\phi_p} \tag{7.20}$$

$$E_c = E_m\left[\frac{3(1+\xi(MRF)\eta_L\phi_p)}{8(1-\eta_L\phi_p)} + \frac{5(1+2(MRF)\eta_T\phi_p)}{8(1-\eta_T\phi_p)}\right] \tag{7.21}$$

7.2.2 *Hui-Shia Model*

Hui-Shia (H–S) model [6, 15] developed the prediction of elastic modulus of unidirectionally aligned two-phase composites with fibre-like or flakes-like filler inclusions. H–S model simplifies the orientation of fillers with the assumption of perfect interfacial bonding between the matrices and fillers with similar Poisson's ratios. The associated elastic modulus equations are given below

Longitudinal elastic modulus (E_{11})

$$\frac{E_c}{E_m} = \frac{E_{11}}{E_m} = \frac{1}{1-(\phi_p/4)\left[\frac{1}{\varepsilon}+\frac{3}{\varepsilon+\Lambda}\right]} \tag{7.22}$$

Transverse elastic modulus (E_{22})

$$\frac{E_c}{E_m} = \frac{E_{22}}{E_m} = \frac{1}{1 - \left[\frac{\phi_p}{\varepsilon}\right]} \tag{7.23}$$

with

$$\varepsilon = \phi_p + \frac{E_m}{E_p - E_m} + 3(1 - \phi_p)\left[\frac{(1-g)\alpha'^2 - (g/2)}{\alpha'^2 - 1}\right] \tag{7.24}$$

$$\Lambda = (1 - \phi_p)\left[\frac{3\left(\alpha'^2 + 0.25\right)g - 2\alpha'^2}{\alpha'^2 - 1}\right] \tag{7.25}$$

where α' is the inverse of aspect ratio (α) or the ratio of thickness to the length of fillers (i.e. $1/\alpha \approx t/l$), and g is a geometric parameter of composites defined in Eqs. 7.24 and 7.25. When $\alpha' \approx 1$, g would be equal to 2/3. The case of $\alpha' \geq 1$ belongs to the inclusion of fibre-like fillers while $\alpha' \leq 1$ is associated with the inclusion of flake-like or disk-like fillers according to Eq. 7.26. In addition, for a perfect interface, g would be given in Eq. 7.27.

$$g = \begin{cases} \frac{\alpha'}{\left(\alpha'^2 - 1\right)^{3/2}}\left[\alpha'\left(\alpha'^2 - 1\right)^{1/2} - cosh^{-1}\alpha'\right] & \alpha' \geq 1 \\ \frac{\alpha'}{\left(1 - \alpha'^2\right)^{3/2}}\left[-\alpha'\left(1 - \alpha'^2\right)^{1/2} + cos^{-1}\alpha'\right] & \alpha' \leq 1 \end{cases} \tag{7.26}$$

$$g = \left(\frac{\pi}{2}\right)\alpha'. \tag{7.27}$$

7.2.3 Laminate Model

Laminate model is developed with respect to geometric properties of filler inclusions in matrices. From the experimental point of view, the orientation of fillers in composites generally contains random orientation or misalignment of fillers as opposed to unidirectionally aligned reinforcements assumed for simplicity in theoretical modelling. The aspect ratio and modulus of fillers, as well as the orientation of dispersed fillers have a significant impact on the elastic modulus of composites [4]. With respect to fully random orientation in the 3D orthogonal direction, platelet fillers would be more effective planar reinforcements as compared with fibre-like fillers. The quantitative equations based on fibre-like or platelet-like fillers are proposed as follows [16, 17]:

$$E_{ran-3D}^{fibre} = 0.184E_{11} + 0.816E_{22} \tag{7.28}$$

$$E_{ran-3D}^{platelet} = 0.49E_{11} + 0.51E_{22} \tag{7.29}$$

where E_{11} and E_{22} are the longitudinal and transverse moduli of composites, respectively.

7.3 Strength of Particulate Polymer Composites

7.3.1 Danusso-Tieghi (D-T) Model

Danusso and Tieghi [18] proposed a good relationship between mechanical strength and volume fraction of rigid matrix based composites. It is assumed in D-T model that there is no adhesion between matrices and fillers, thus without load transfer from the matrices to fillers. This means that the total load is equal to that carried by the matrices alone. In addition, D-T model can be applied from regular or irregular filler distribution to random distribution. Nonetheless, this model can also be used when the adhesion occurs. A simple equation for the strength of composites is given by:

$$\sigma_c = \sigma_m(1 - \psi) \tag{7.30}$$

where σ_c and σ_m are the tensile strengths of composites and matrices, respectively and ψ is the area fraction in the cross-section. With those above-mentioned model assumptions and random orientation of filler structures, the area fraction of matrix must be the same for any cross-section through the matrix, for which the area fraction of the matrix cross-section is equal to the volume fraction of fillers ($\psi \approx \phi_p$) [18, 19].

7.3.2 Nicolais and Narkis (N–N) Model

This model is based on previous D-T model by replacing the volume fraction with a power law function in terms of volume fraction as follows [20]

$$\sigma_c = \sigma_m\left(1 - a\phi_p^b\right) \tag{7.31}$$

where a and b are constants that are influenced by particle shape and arrangement in composites. Based on the hypothesis that there is no adhesion between matrices and fillers, N–N model is considered for cubic filler shape with uniformly dispersed spherical particles. However, it is assumed that a minimum cross-section of continuous phase should be perpendicular to the applied load. As such, the strength of composites is calculated based on the equation below

$$\sigma_c = \sigma_m\left(1 - 1.21\phi_p^{2/3}\right) \tag{7.32}$$

It is used to predict the lower-bound strength of composites. In addition, for their upper bound, the strength is calculated with the assumption of perfect adhesion between matrices and fillers. Accordingly, the strength of composites is the same as that the strength of matrices or unfilled polymers ($\sigma_c \approx \sigma_m$).

7.3.3 *Lu Model*

Lu model [21] is developed by modifying N–N model when there is some adhesion existing between matrices and fillers, which is given by

$$\sigma_c = \sigma_m\left(1 - 1.07\phi_p^{2/3}\right) \tag{7.33}$$

The equation is established by using the combined properties from micromechanical measurement, microdamage monitoring, as well as micromechanical analysis in order to obtain the interfacial bonding strength between matrices and fillers.

7.3.4 *Turcsányi-Pukànszky-Tüdõs (T-P-T) Model*

Turcsányi et al. [22] proposed a semi-empirical equation for very strong particle–matrix interfacial bonding with a simple hyperbolic function to describe the change of filler cross-section with the filler content. Nonetheless, the functionalisation of the exponential function is subjected to the unexpected dependence of the strength of matrices and volume fraction of fillers. Additionally, T-P-T model can be applied to composites with the inclusion of spherical particles, as well as anisotropic particles. The equation is generally given by

$$\sigma_c = \left[\frac{1 - \phi_p}{1 + 2.5\phi_p}\sigma_m\right]\exp\left(B\phi_p\right) \tag{7.34}$$

where σ_m is the strength of matrices. B is a constant that depends on the surface area of particles, their density and interfacial bonding with matrices. If $B = 0$, the fillers act as the voids having poor interfacial bonding without adhesion and load transfer at the matrix-filler interface. Nonetheless, if $B \leq 3$, the filler-matrices interface is weak without reinforcing effect. The constant B was calculated by fitting experimental data with theoretical results and derived from the minimum sum of squares of variance from experimental data of composite strengths.

In the following section, the theoretically predicted tensile properties (i.e. tensile modulus and tensile strength) according to Eqs. (7.1)–(7.34) were compared with experimental data obtained from the tensile tests of bioepoxy/clay nanocomposites cured by isophorone diamine (IPDA). The discussion was separated into two categories in relation to the effects of clay content and ESO content on tensile properties of bionanocomposites.

7.4 Tensile Modulus of Bionanocomposites

7.4.1 *Effect of Clay Content*

ROM and modified ROM based models according to Eqs. (7.1)–(7.29) were used to predict the effect of clay content on tensile moduli of bioepoxy/clay nanocomposites cured by IPDA. Material properties of bionanocomposite constituents including moduli of single platelet MMT (E_{MMT}), DGEBA conventional epoxy resin, ESO as well as Poisson's ratio and density of matrices (i.e. DGEBA resin and ESO) and Cloisite 15 nanofillers are listed in Table 2.1. In addition, Eq. 7.6 was used to calculate the effective elastic modulus of clay intercalated structures taking place in bionanocomposites. The *d*-spacing values of clay intercalated structures were derived from XRD results. The calculation results for the effective elastic modulus of clay intercalated structures in bionanocomposites are listed in Table 7.1. The effective elastic modulus of clay intercalated structures was used to calculate theoretical elastic moduli of bionanocomposites at different clay contents. Additionally, with respect to the random orientation of clay structures, 3D platelet-like filler laminate model was employed to incorporate actual filler geometries. Nonetheless, MRF $\approx$ 0.66 was also used, whose assumption was based on the combination of partially intercalated/exfoliated and aggregated clay structures in bionanocomposites. The above-mentioned geometric parameters including clay orientation and dispersion status were determined according to TEM micrographs of bionanocomposites based on IPDA at the fixed ESO content of 20 wt%, as shown in Fig. 5.8.

Theoretical prediction of elastic moduli of bioepoxy/clay nanocomposites as a function of clay content was compared with experimental data, illustrated in Figs. 7.1, 7.2 and 7.3. It can be seen in Fig. 7.1 that Voigt model demonstrated an overestimated result as opposed to experimental data despite the inclusion of MRF. On the contrary, the theoretical prediction of elastic modulus by Reuss model, which is known as inverse Voigt model, underestimates the experimental data. The overall elastic moduli predicted by Voigt model and Reuss model offer a rough estimation for elastic moduli of bionanocomposites, thus demonstrating an extreme condition for upper and lower bounds of elastic moduli, respectively. This phenomenon indicated that those two models neglected geometry, dispersion status and orientation of fillers, which further made the theoretical prediction become questionable.

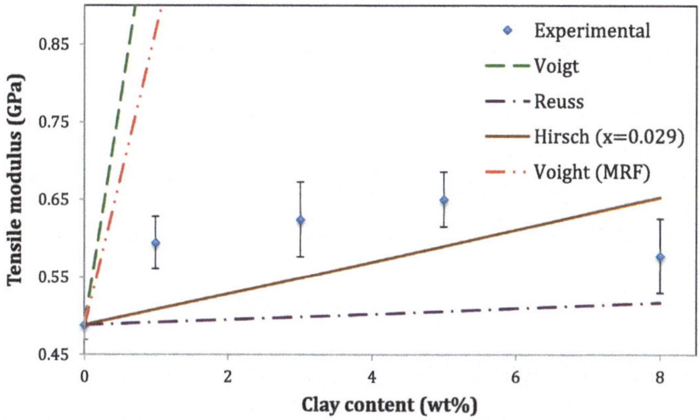

Fig. 7.1 Comparison between theoretical prediction and experimental data for elastic moduli of bioepoxy/clay nanocomposites as a function of clay content [23]

In order to enhance the prediction accuracy, filler geometric parameters, filler dispersion status, as well as filler orientation were taken into account by applying Hirsch model and Halpin–Tsai models. It is assumed in Hirsch model that the random orientation of clay fillers takes place with the combination of parallel and series filler orientations in bionanocomposites, in which "x" is regarded as a significant parameter in predicting elastic moduli of bionanocomposites. As for "x" parameter in Hirsch model, it can also be determined that filler orientation is more influenced by elastic moduli of bionanocomposites. The curve fitting with experimental data by Hirsch model revealed better agreement between Hirsch model prediction and experimental data, as compared with those predicted by Voigt model and Reuss model. Nonetheless, the value of "x" was determined to be 0.029 for parallel filler orientation to the stress direction. This may suggest that the series orientation of clay platelet fillers was significantly affected by the behaviour of bionanocomposites in real experiments.

On the other hand, in order to enhance the accuracy of geometric parameters, H-T model and modified H-T model were employed in predicting elastic moduli of bionanocomposites by varying aspect ratios of fillers. Elastic modulus prediction of bionanocomposites was obtained according to Eqs. 7.12–7.21 in the longitudinal and transverse directions of original H-T model equations, as well as modified equations with the incorporation of MRF. Nonetheless, since the random orientation of clay fillers was observed based on TEM observation, the 3D-platelet laminate model for random orientation (Eq. 7.29) was employed in predicting elastic moduli of bionanocomposites as a function of clay volume fraction (ESO content: 20 wt%). Figure 7.2 presented the theoretical prediction and experimental data for elastic moduli of bionanocomposites obtained by H-T model and modified H-T models with laminate model orientation. Figure 7.2a showed that the increase in aspect ratio from 3 to 60, leading to the enhanced wettability and contact surface areas between

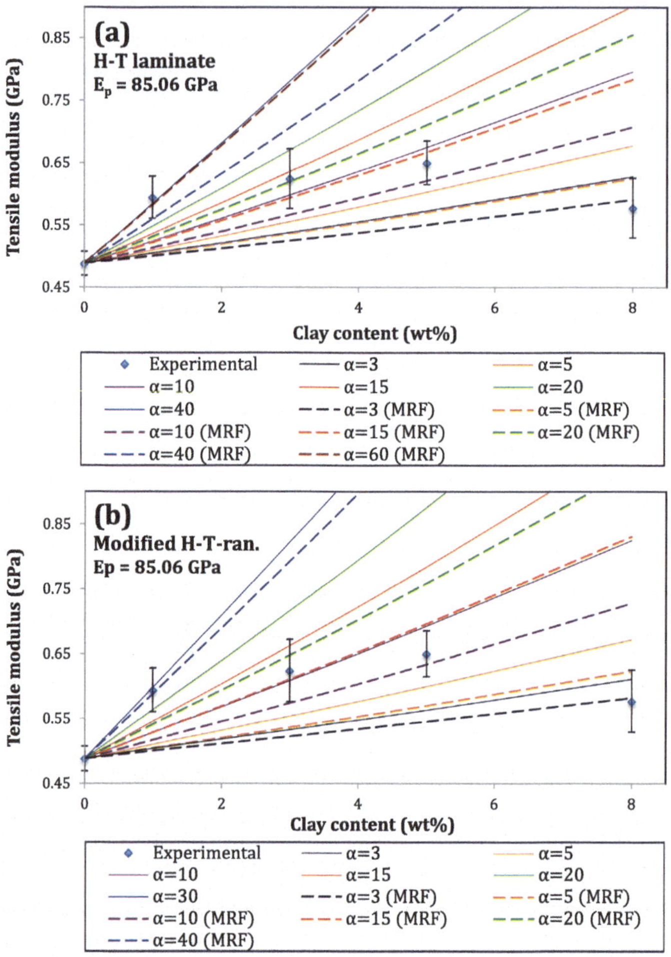

Fig. 7.2 Theoretical prediction curves for elastic moduli of bionanocomposites with intercalated and randomly oriented clay fillers: **a** H-T laminate model and **b** modified H-T model with random orientation [23]

clay fillers and polymer matrices, demonstrated an improvement of estimated elastic moduli in line with increasing the clay content. It is manifested that increasing the clay content may result in various aspect ratios for clay fillers in matrices, which is associated with different clay wettability in bionanocomposites. The lower clay content yields higher filler aspect ratios as expected. Nonetheless, the use of MRF in H-T model with laminate model based on random orientation exhibited a reduction of predicted elastic moduli, which could eventually affect the estimation of aspect ratios for clay fillers. The percentage difference between experimental data and theoretical results was ranging from 2–8.7% and 0.7–2.7% for H-T laminate model and H-T laminate model with MRF, respectively.

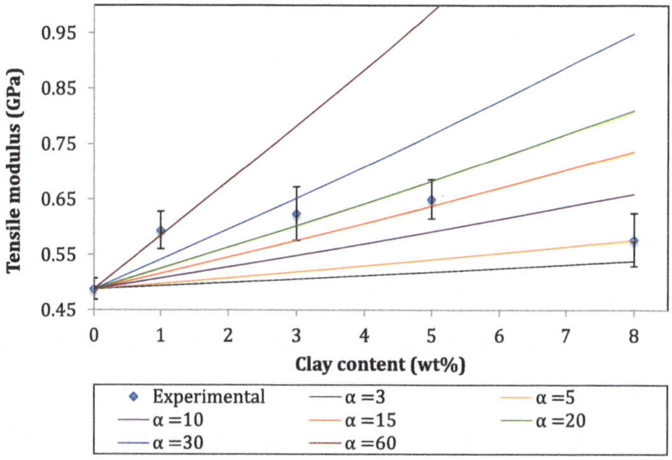

Fig. 7.3 Comparison between theoretical prediction and experimental data based on H–S laminate model at different filler aspect ratios [23]

Similar approaches were also revealed for the theoretical prediction of elastic moduli of nanocomposites based on modified H-T model with the random orientation of clay fillers (Eq. 7.16), as shown in Fig. 7.2b. The predicted elastic moduli of bionanocomposites offered a fairly close agreement with experimental data. The percentage errors for modified H-T model with the random filler orientation was in range of 0.7–7.6%. On the other hand, for those model with the incorporation of MRF, the percentage errors were found to be between 0.9 and 2.5%. Detail percentage errors of each theoretical model can be found in Table 7.2. Consequently, the prediction of

Table 7.2 Percentage errors in theoretical models in comparison with experiment data at different clay contents in bionanocomposites [23]

Theoretical model	Percentage error (%)			
Clay content (wt%)	1	3	5	8
Voigt	79.54	259.82	428.63	809.72
Voigt (MRF)	46.25	163.54	273.55	527.57
Reuss	−17.25	−20.11	−22.19	−10.41
Hirsch	−14.47	−12.09	−9.27	13.10
H-T laminate random	2.75	1.95	3.27	8.66
H-T laminate random with MRF	−1.98	−0.74	2.73	2.28
Modified H-T random	0.71	−2.40	6.59	5.85
Modified H-T random with MRF	−0.91	−2.03	−2.46	0.85
H–S laminate	−1.60	−3.48	−1.84	−0.07

Note Negative values in percentage errors represent lower prediction values than experimental data

elastic modulus by H-T laminate model and modified H-T-random model with MRF appeared to be closer to experimental data relative to those without MRF.

Additionally, elastic moduli of bionanocomposites were also predicted by Hui-Shia (H–S) laminate model, as shown in Fig. 7.3, in terms of aspect ratio of clay fillers and clay content. Similarly, with the implementation of H-T model and modified H-T model, theoretically estimated elastic moduli of bionanocomposites exhibited different aspect ratios with increasing the clay content. The increase in clay content revealed the reduction in aspect ratios of clay fillers in bionanocomposites. Nonetheless, the percentage errors between theoretical results and experimental data were ranging from 0.07 to 3.5%, indicating their good correlation with experimental data. It is implied that H–S-laminate model offers a better theoretical prediction for elastic moduli of bionanocomposites as opposed to other theoretical models such as ROM model, Hirsch model, H-T laminate model and modified H-T-random model.

Clay aspect ratio revealed special geometric characteristic of clay fillers in bionanocomposites. When clay platelet layers are dispersed well in matrices with the existence of single platelet layers, this leads to exfoliated structures with the aspect ratio as high as 1000 as the benchmark [24]. However, with respect to undispersed clay fillers in matrices, their aspect ratios can be significantly reduced to approximately 10. The theoretical prediction of elastic modulus by H-T model and H–S model could obtain the estimated aspect ratios for clay fillers in bionanocomposites systems at different clay contents, as shown in Table 7.3. It could be observed that at a lower clay content of 1 wt%, effective aspect ratio reached the highest value in the range of 30- 60 though it was still lower than that of single clay platelet. When the clay content increased up to 3 and 5 wt%, effective aspect ratios decreased to a range between 15–30 and 10–15, respectively. In addition, experiment data at the highest clay content of 8 wt% revealed the lowest aspect ratio ($\alpha \leq 5$), indicating poor clay dispersion in nanocomposites. In relation to effective aspect ratios of clay particles, it was noted that a close aspect ratio prediction at clay content from 3 to 8 wt% took place in H-T-laminate model and modified H-T-random model with MRF. Similarly, a close aspect ratio prediction was also detected in both H-T models without MRF at similar clay contents (3 to 8 wt%). However, at the lower clay content of 1 wt%, the prediction of various aspect ratios was obtained for different H-T models, which

Table 7.3 Aspect ratios of clay fillers based on curve fitting of experimental data and theoretical prediction in bionancomposites [23]

Theoretical model	Aspect ratio			
Clay content (wt%)	1	3	5	8
H-T laminate	~40	10–15	5–10	<3
H-T laminate with MRF	~60	~20	10–15	<3
Modified H-T random	~30	10–15	5–10	<3
Modified H-T random with MRF	~40	15–20	10–15	~3
H–S laminate	~60	20–30	~15	~5

could be attributed to diverse assumptions used for determining geometric parameters to predict elastic moduli of nanocomposites based on H-T laminate model and modified random oriented model.

On the other hand, H–S model exhibited higher aspect ratio prediction with a more sensitivity when compared with H-T models, as shown in Table 7.3. The aspect ratios predicted by H–S laminate model appeared to be 50% higher than those based on H-T models. Despite different aspect ratios, the prediction of effective aspect ratios is consistent with morphological structures of bionanocomposites depicted in Fig. 5.8, in presence of random orientation of clay fillers in matrices, as well as an increasing level of clay aggregation when increasing the clay content. More interestingly, elastic modulus predictions obtained by H–S laminate model were closer to experimental data as opposed to those by H-T models, especially at clay contents from 5 to 8 wt%, given in Table 7.2. The percentage errors of predicted elastic modulus of bionanocomposites at the clay contents of 5 and 8 wt% by H–S laminate model were found to be approximately 1.84 and 0.07%, respectively. Nonetheless, those predicted by modified H-T-random model were about 2.46 and 0.85% accordingly. In contrast, at lower clay contents of 1 and 3 wt%, the percentage errors of predicted elastic moduli by H–S laminate model were higher than those of H-T models (i.e. about 1.6 and 3.48%, respectively). As such, it is implied that H–S laminate random oriented model is more applicable to predict elastic moduli of bionanocomposites beyond the clay content of 5 wt%. On the other hand, H-T laminate model and modified H-T random oriented model with/without MRF were confirmed to be more suitable for nanocomposites at the clay content below 3 wt%.

7.4.2 Effect of ESO Content

Similar to the effect of clay content, theoretical predictions to elastic moduli of bionanocomposites were compared with experimental data. Equations (7.1)–(7.29) were also employed to predict elastic moduli of bionanocomposites in terms of bioepoxy blend matrix content and corresponding estimated results were depicted in Figs. 7.4, 7.5, 7.6 and 7.7. At different ESO contents from 0 to 60 wt%, elastic moduli of bioepoxy blends varied according to the ESO contents used, as shown in Table 7.4. In a similar manner, the densities of blend matrices were also altered based on different ESO contents. Additionally, the dispersion status of clay fillers affected elastic moduli of bionanocomposites. Hence elastic moduli of clay particles were calculated based on Eq. (7.6) along with the results listed in Table 7.4. Such elastic moduli of clay particles were calculated with the assumption that intercalated clay structures were dominantly formed as compared to other dispersion statuses (i.e. exfoliated and aggregated clay structures), which were based on XRD analysis results shown in Fig. 5.6a. Other parameters used in those theoretical models were also summarised in Table 2.1. The clay content of 5 wt% was fixed at all different ESO contents in bionanocomposites.

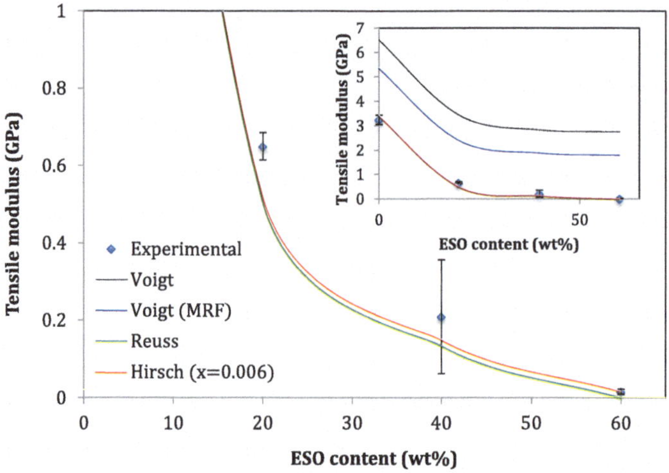

Fig. 7.4 Comparison between theoretical prediction and experimental data for elastic moduli of bionanocomposites as a function of ESO content [23]

Table 7.4 Tensile properties of blend matrices and effective moduli of intercalated clay particles in bionanocomposites (fixed clay content: 5 wt%) [23]

ESO content	E_{blend}	σ_{blend}	ρ_{blend}	Bionanocomposites reinforced with fixed 5 wt% clays	
(wt%)	GPa	MPa	$g{\cdot}cm^{-3}$	d_{002} (nm)	$E_{intercalation}$ (GPa)
0	3.30	59.56	1.17	1.80	92.72
20	0.49	28.53	1.13	1.96	85.59
40	0.13	10.26	1.09	2.07	80.65
60	1×10^{-4}	0.29	1.06	1.97	85.08

In Fig. 7.4, elastic moduli of bionanocomposites showed a declining trend with increasing the ESO content. As previously mentioned, the decrease in elastic modulus could result from ESO plasticisation effect. Increasing the ESO content yielded bionanocomposites with more rubber-like behaviour and less brittle nature in material properties. Voigt model and Reuss model could only provide rough predictions for elastic moduli of bionanocomposites at different ESO contents, as shown in Fig. 7.4. It was seen that theoretical prediction using Reuss model as the lower bound led to better estimation than Voigt model as the upper bound. This result suggested that series model for filler-matrix orientation became more dominant when compared with parallel model. Additionally, since the orientation of clay fillers in matrices was observed to be more randomly distributed, the combination of series and parallel models could be applied in predicting elastic moduli of nanocomposites by using Hirsch model. The curve fitting with experimental data revealed the controlling factor "x" was determined to be about 0.006 for clay nanocomposites in parallel orientation.

Due to the "x" value between 0 and 1, the controlling factor for series model was found to be equal to 0.994. This might suggest that the series model for fibre-matrix orientation became more prevalent. Such a finding had good agreement with Reuss model as mentioned earlier.

Additionally, it was evidently showed from Fig. 7.4 that there was a clear difference between experimental results and theoretical modelling prediction. Calculated percentage errors for each theoretical model were listed in Table 7.5, which were obtained from the percentage of the difference between theoretical results and experimental data. As clearly seen, theoretical predictions using Reuss model and Hirsch model became much closer to experimental data as opposed to Voigt model and Voigt model with MRF. For instance, without ESO (i.e. 0 wt% ESO), the percentage errors of Reuss model and Hirsch model exhibited comparable values of approximately 5.30 and 5.89%, respectively, with reference to Voigt model having about a percentage error of 100%. More interestingly, theoretical predictions of elastic moduli of bionanocomposites with 60 wt% ESO were shown to coincide with experimental data using Hirsch model. The percentage error of proposed Hirsch model at the ESO content of 60 wt% was approximately 7%, which was the closer prediction when compared with other theoretical models presented in Table 7.5.

With respect to the random orientation of clay fillers in matrices, 3D laminate model was employed to predict elastic moduli of bionanocomposites at different ESO contents, along with H-T model and H–S model according to Eqs. (7.12) to (7.29). Nonetheless, since it was found that the clay structures were the combination of partially clay intercalation/exfoliation, as well as clay aggregation, the MRF equal to 0.66 was applied to H-T model equations. The curves showing the elastic modulus prediction by H-T laminate model and modified H-T random oriented model with MRF were depicted in Figs. 7.5 and 7.6. H-T laminate model with MRF in Fig. 7.5 was in good accordance with experimental data of bionanocomposites especially at

Table 7.5 Percentage errors in theoretical predictions in comparison with experimental data based on different theoretical models at various ESO contents in bionancomposites [23]

Theoretical model	Percentage error (%)			
ESO content (wt%)	0	20	40	60
Voigt	100.08	428.63	1249.97	17,121.94
Voigt (MRF)	65.38	273.55	811.35	11,266.69
Reuss	5.30	−22.19	−35.87	−99.35
Hirsch	5.89	−9.27	−27.91	7.20
H-T laminate	22.48	3.27	5.64	−98.52
H-T laminate with MRF	17.39	2.73	5.24	−98.80
Modified H-T random	16.04	6.59	−1.20	−98.90
Modified H-T random with MRF	12.38	−2.46	−5.35	−99.06
H–S laminate random	7.79	−1.84	−1.16	−99.01

Note Negative values in percentage error represent lower prediction values than experimental data

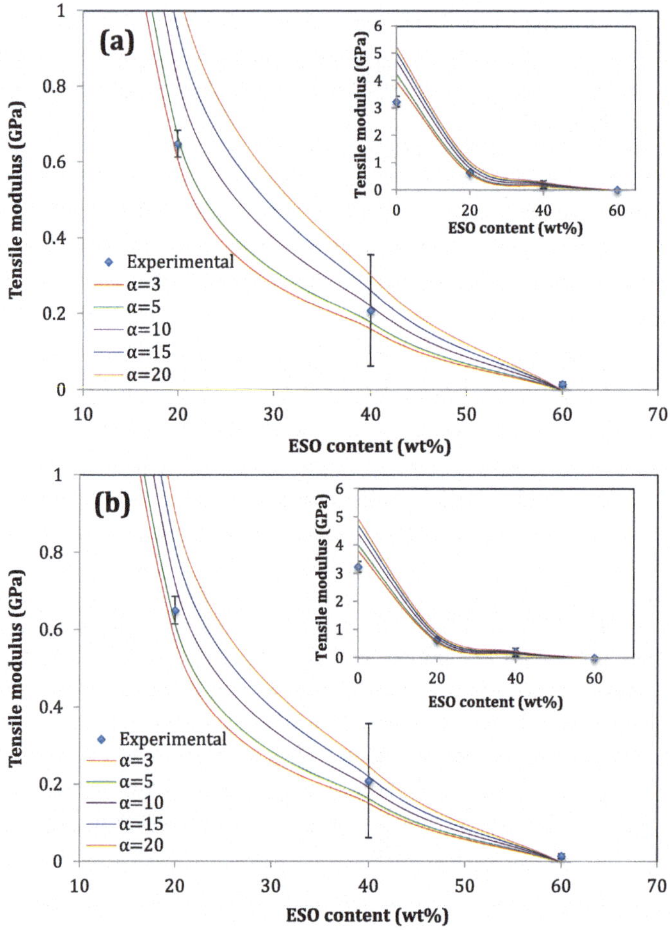

Fig. 7.5 Comparison between theoretical prediction and experimental data for elastic moduli of bionanocomposites using different aspect ratios calculated based on **a** H-T laminate model and **b** H-T laminate model with MRF [23]

the ESO contents of 20 and 40 wt%. This led to a small percentage of the difference in a range from 3 to 5.6% accordingly. In addition, at the ESO content of 0 wt%, the theoretical prediction appeared to overestimate experimental data as opposed to their underestimation at a higher ESO content of 60 wt%. The overestimated value could be attributed to geometric constants involved in the equations, as detected from theoretical prediction of Reuss model and Hirsch model ($x = 0.006$). Those findings infer that a dominantly transverse orientation of clay structures in conventional epoxy matrices has been formed.

On the other hand, modified H-T random model according to Eq. (7.16) and modified H-T random model with MRF based on Eq. (7.21) were demonstrated in Fig. 7.6. It was seen that the modification of H-T model affected the prediction of

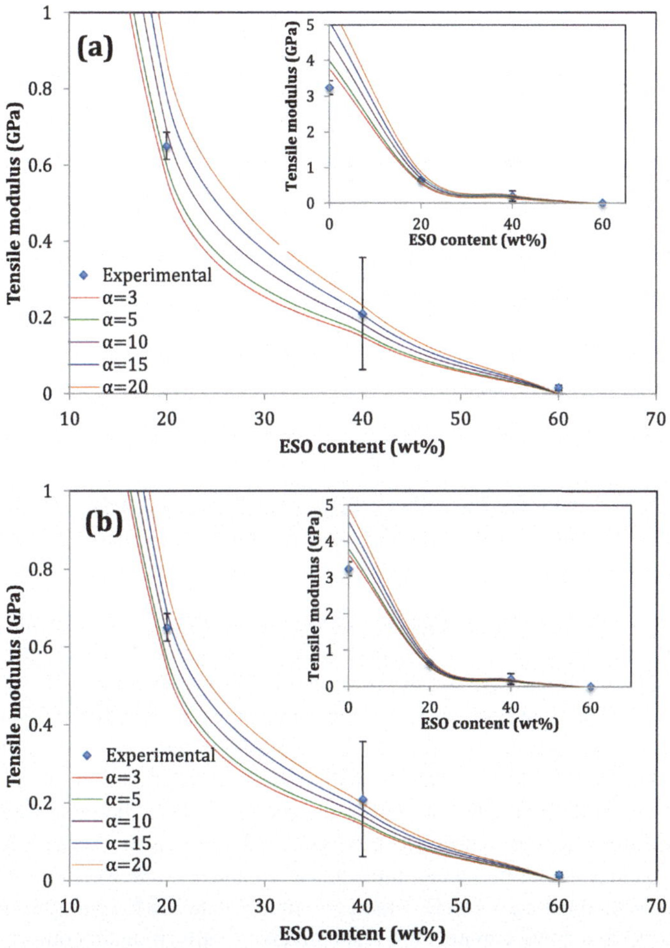

Fig. 7.6 Comparison between theoretical prediction and experimental data for elastic moduli of bionanocomposites at different aspect ratios calculated based on **a** modified H-T random oriented model and **b** modified H-T random oriented model with MRF [23]

filler aspect ratios, especially at the ESO contents of 20 and 40 wt%, even though the percentage errors of modified H-T random model with MRF were comparable to those of H-T laminate model with MRF, as shown in Table 7.5. Additionally, the theoretical modulus prediction by using H-T laminate model and modified H-T random oriented model could be achieved by estimating the aspect ratios of clay fillers in bionanocomposites at different ESO contents. It was found that filler aspect ratios of bionanocomposites varied according to various ESO contents, as shown in Fig. 7.5. Nonetheless, the use of MRF showed an increasing aspect ratio value as compared with the counterpart without MRF. The aspect ratios of conventional nanocomposites (i.e. 0 wt% ESO) exhibited comparable values when predicted using

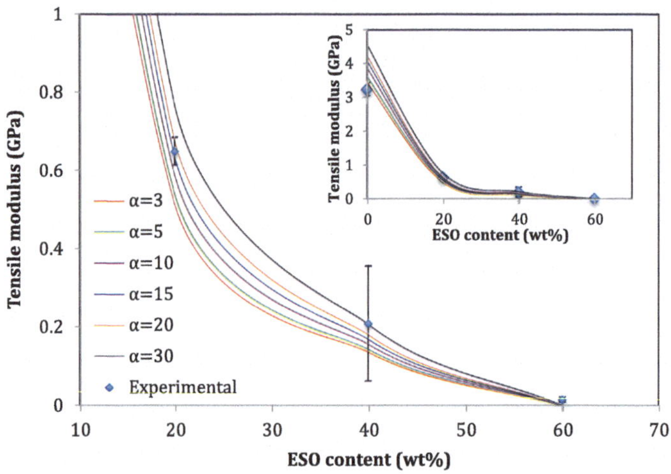

Fig. 7.7 Elastic modulus predictions of bionanocomposites reinforced with 5 wt% Cloisite 15 clays as a function of ESO content for different filler aspect ratios using H–S laminate model [23]

H-T laminate model and modified H-T random oriented model with the incorporation of MRF, as evidenced by a lower aspect ratio value below 3. This could suggest that clay fillers were undispersed in conventional epoxy, which was likely to form aggregated clay fillers in the matrices. However, a further increase in ESO content revealed increasing aspect ratio values. At the ESO content of 20 wt%, the aspect ratio predicted by H-T laminate model with MRF was improved in range of 5–10. In comparison, additional ESO at approximately 40 wt% yielded the aspect ratios of 10–15. Moreover, with respect to modified H-T random model with MRF, the predicted aspect ratios of bionanocomposites at 20 and 40 wt% were slightly higher as compared with those based on H-T laminate model with MRF, leading to the aspect ratios of 10–15 and approximately 20, respectively. Finally, bionanocomposites with 60 wt% ESO exhibited an aspect ratio over 20 for both H-T model and modified H-T model. Nonetheless, the percentage errors of each H-T model appeared to be similar, which were ~ 99% less than those obtained from corresponding experimental data. This underestimation of elastic modulus could be attributed to particle size effect in relation to filler aspect ratio and particle content [4, 25]. Since aspect ratio is indicative of clay particle size dispersed in bioepoxy matrices, the reduction of clay particle size would inevitably increase the aspect ratios of clay fillers. Similarly, with the effect of filler inclusion, the increase in filler aspect ratio would reduce the size of dispersed clay fillers in the matrices as expected, which might further improve the rigidity of nanocomposites. More interestingly, the high elastic modulus of bionanocomposites at 85.08 GPa in experimental data with 60 wt% ESO was likely to be associated with a significantly higher modulus of clay particles as opposed to that of corresponding bioepoxy (i.e. 10^{-4} GPa). The high elastic modulus of clay fillers could improve the stiffness of bionanocomposites. Direct effect from the clay inclusion into rubber-like bioepoxy matrices would greatly affect the deformation reduction of bionanocomposites as opposed to neat bioepoxy at high ESO contents.

In addition to the effect of aspect ratio and filler modulus, the orientation of dispersed clay fillers has also significantly impacted elastic modulus, as shown from the percentage of the difference between theoretical prediction calculated by Eq. (7.29) for H-T laminate-random model and Eq. (7.16) for modified H-T random model and experimental data. In general, those two equations are established from similar theoretical elastic moduli in H-T model, namely longitudinal (E_{11}) modulus and transverse modulus (E_{22}). However, diverse assumptions for the interpretation of geometry direction revealed different results for the prediction of elastic moduli of nanocomposites. From experimental point of view, modified H-T random model with lower percentage errors, is more applicable to predict elastic moduli of bionanocomposites as compared with H-T laminate random oriented model. This can be associated with different explanation for determining the constants of longitudinal direction (E_{11}) variable, which are 0.375 and 0.49 for modified H-T random oriented model and H-T laminate model, respectively. Furthermore, it is very critical to predict filler aspect ratios in the matrices based on the theoretical prediction in relation to elastic moduli of nanocomposites with the combination of random orientation of fillers in polymer matrices.

On the other hand, with respect to H–S laminate model, the theoretical prediction of bionanocomposites compared to experimental data was shown in Fig. 7.7. Clearly, curve fitting by H–S laminate model demonstrated a good prediction for filler aspect ratio in experiment data. The aspect ratios predicted by H–S laminate model were summarised in Table 7.6. Increasing the ESO content dramatically enhanced the aspect ratios of clay fillers in bionanocomposites. Conventional nanocomposites (i.e. 0 wt% ESO) presented the lowest aspect ratios below 3. The aspect ratios of bionanocomposites with 20 wt% ESO were in range of 15–20, as opposed to nearly 30 for bionanocomposites at the ESO content of 40 wt%. When the ESO content reached 60 wt%, the predicted filler aspect ratio reached a level higher than 30, as illustrated in Fig. 7.7. Additionally, the effect of clay content also demonstrated that the theoretical prediction of elastic moduli of bionanocomposites determined by H–S laminate model demonstrated closer prediction values as shown in Table 7.5. The percentage errors for bionanocomposites up to 40 wt% ESO was below 7%,

Table 7.6 Aspect ratios of clay fillers based on curve fitting of experimental data with theoretical models at different ESO contents in bionancomposites [23]

Theoretical model	Aspect ratio			
ESO content (wt%)	0	20	40	60
H-T laminate	<3	3–5	~10	>20
H-T laminate with MRF	<3	5–10	10–15	>20
Modified H-T random	<3	5–10	~15	>20
Modified H-T random with MRF	<3	10–15	~20	>20
H–S laminate	<3	15–20	~30	>30

which appeared to be greater up to almost 100% at the ESO content of 60 wt%. When compared with other theoretical models, the application of H–S laminate model yielded closer elastic modulus prediction with much lower percentage errors, especially for bionanocomposites at the ESO contents of 20 and 40 wt% (i.e. ~ 1.84 and 1.16%, respectively). Moreover, filler aspect ratios predicted by H–S laminate model remained at a high level, as compared to those estimated by other theoretical models. As a result, when the effect of ESO content on bionanocomposites reinforced by 5 wt% clays was investigated, it was evident that H–S laminate model appeared to be more reliable in predicting their elastic moduli in experimental data. This is especially the case for bionanocomposites at the ESO contents below 40 wt%. On the other hand, when the ESO content is as high as 60 wt%, Hirsch model with the combination of parallel and series orientations became more applicable in estimating elastic moduli of bionanocomposites.

7.5 Tensile Strength of Bionanocomposites

7.5.1 Effect of Clay Fillers

The theoretical tensile strength values predicted by Eqs. (7.30) to (7.34) were compared with the experimental data of bionanocomposites cured by IPDA at a fixed ESO content of 20 wt%. Figure 7.8 showed the theoretical prediction of nanocomposites by using different theoretical models, which were applied to fit experimental data. Tensile strengths of conventional epoxy/clay nanocomposites demonstrated a declining strength trend with increasing the clay content from 1 to 8 wt%, as shown in Fig. 7.8a. It was observed that a comparable correlation between theoretical results and experimental data predicted by Nicolais and Narkis (N–N) model and Turcsányi-Pukànszky-Tüdõs (T-P-T) model was detected when compared with Lu model and Danusso-Tieghi (D-T) model. Lu model is adapted from modified N–N model with the power law function of filler volume fraction. Power-law constants used were determined to be 1.07 and 0.67, representing particle shape and orientation in nanocomposites, respectively. Furthermore, it was seen in Fig. 7.8a that the theoretical prediction of tensile strength by using Lu model was comparably equal to the strength of epoxy matrices. With respect to D-T model, this model demonstrated a comparable agreement on the tensile strength of conventional nanocomposites at the low clay content of 1 wt%. Conversely, the prediction failed at higher clay contents, as evidenced in Fig. 7.8a.

In the case of T-P-T model, Eq. (7.34) has been used to predict the tensile strengths of conventional nanocomposites as a function of volume fraction of clay fillers illustrated in Fig. 7.8a. Two different B parameters are used in this model, in which $B = 0$ for bionanocomposites with poor interfacial bonding between clay fillers and matrices where B is determined from the calculation based on the least square method [22] in curve fitting. The curve fitting method with experimental data from

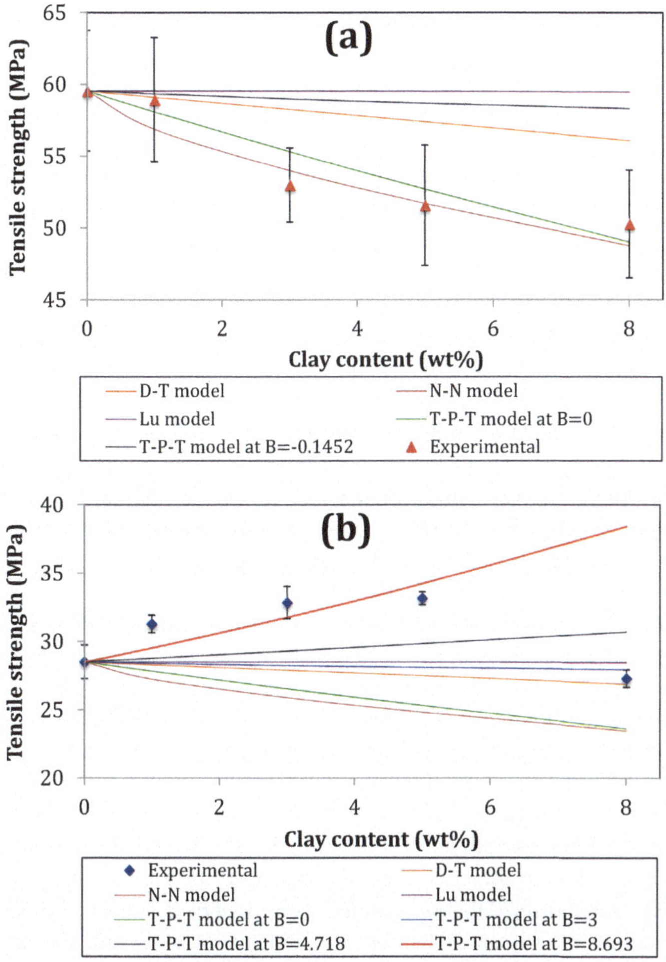

Fig. 7.8 Comparison between theoretical prediction and experimental data of tensile strengths of nanocomposites at different clay contents: **a** Conventional epoxy/clay nanocomposites and **b** bioepoxy/clay nanocomposites [23]

0 to 60 wt% ESO by T-P-T model yields the empirical constant of experimental data, which is $B = -0.145$. When this value is assigned with $B = 0$, it indicated the poor interfacial bonding took place between clay fillers and matrices in nanocomposites. In addition to T-P-T model, N–N model was also applied to confirm the finding from the prediction of T-P-T model. This model deals with a power-law relation with the simple geometry, presenting a lower-bound theoretical prediction for the tensile strengths of conventional nanocomposites. As shown in Fig. 7.8a, the theoretical prediction trends and values of tensile strengths by N–N model were

approximately the same as experimental data. As such, experimental data for conventional epoxy/clay nanocomposites implied there was no adhesion occurring between clay fillers and epoxy matrices according to this model.

On the other hand, the theoretical prediction for tensile strengths of bioepoxy/clay nanocomposites at the fixed ESO content of 20 wt% was revealed in Fig. 7.8b. Similar to the theoretical prediction of conventional nanocomposites, four different theoretical models were also applied to predict the tensile strength values for bionanocomposites. Theoretical predictions based on N–N model and T-P-T model ($B = 0$) underestimated experimental data for tensile strengths of bionanocomposites. This indicated that bionanocomposites possessed better interfacial adhesion between clay fillers and matrices, which demonstrated an increase in tensile strength with the inclusion of clay fillers. The further analysis for experimental data of bionanocomposites was based on the adoption of D-T model and Lu model to experiments data. As shown in Fig. 7.8b, D-T model and T-P-T model ($B = 3$) appeared to be closer to experiment data at the high clay content of 8 wt%. Furthermore, such results suggested that those bionanocomposites might have poor interfacial bonding. Nonetheless, this prediction had good agreement with scanning electron microscopic (SEM) analysis results, as shown in Fig. 5.12. The SEM images presented existing microcracks close to interfacial areas between clay fillers and matrices.

Additionally, tensile strength of bionanocomposites with 20 wt% ESO was also predicted by T-P-T model, as shown in Fig. 7.8b. Similar to the use of T-P-T model in experimental data for conventional nanocomposites, the curve fitting by using T-P-T model made B value equal to 4.718. This B value can suggest that bionanocomposites benefit from reinforcing effect with increasing the clay fillers and better matrix-filler interfacial bonding when B is over 3. Moreover, due to the great difference between theoretical prediction and experimental data in bionanocomposites at the clay content of 8 wt%, the theoretical prediction by T-P-T model was also conducted for experimental data at the clay contents ranging from 1 to 5 wt%, as shown in Fig. 7.4b. It was determined that the empirical constant of B was equal to 8.693. This acquisition of B value indicated that bionanocomposites at clay contents ranging from 1 to 5 wt% exhibited better interfacial bonding, with a clear sign of more efficient filler-matrix load transfer when compared with those at the clay content of 8 wt%.

7.5.2 *Effect of ESO Content*

Similar to the effect of clay content, the prediction for tensile strength of bionanocomposites at a fixed clay content of 5 wt% was compared with experimental data as a function of ESO content, Fig. 7.9. It was clearly revealed that experimental data for tensile strengths of bionanocomposites were reduced with increasing the ESO content from 20 to 60 wt% despite the strength improvement as opposed to that of neat matrices. In addition, the tensile strengths of conventional nanocomposites (i.e. 0 wt% ESO) exhibited a declining tendency with additional 5 wt% clay fillers.

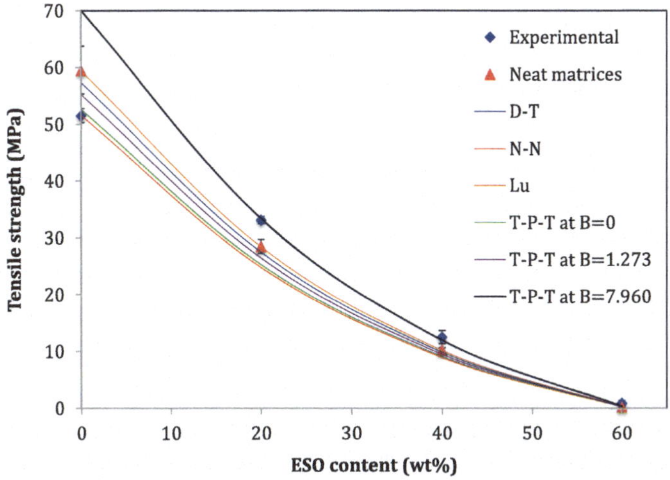

Fig. 7.9 Tensile strength predictions of bionanocomposites reinforced with 5 wt% Cloisite 15 clays as a function of ESO content using different theoretical models [23]

With respect to theoretical prediction, as can be seen in Fig. 7.9, all proposed models did not demonstrate good predictions to fit all experimental data of tensile strengths. The percentage errors of proposed model prediction for tensile strength were enhanced by increasing the ESO content, as shown in Table 7.7. The proposed models could thus only fit one or two experimental data. From the hypothesis of particle–matrix interfacial bonding, N–N model and T-P-T model ($B = 0$) were much closer to experimental data for conventional nanocomposites (0 wt% ESO). The percentage errors of N–N model and T-P-T model ($B = 0$) yielded comparable values of approximately 0.27 and 2.17%, respectively, as compared to experimental data (0 wt% ESO). Since N–N model and T-P-T model ($B = 0$) were categorised as lower bounds for tensile strength models, it could be suggested that conventional

Table 7.7 Percentage errors for proposed tensile strength predictions of bionanocomposites at different ESO contents [23]

Theoretical model	Percentage error (%)			
ESO content (wt%)	0	20	40	60
D-T model	11.31	−16.98	−20.87	−70.02
N–N model	0.27	−25.07	−28.43	−72.83
Lu model	15.38	−14.04	−18.16	−69.03
T-P-T model (B = 0)	2.17	−23.60	−26.98	−72.46
T-P-T model (B = 1.273)	7.21	−20.13	−23.95	−71.42
T-P-T model (B = 7.960)	35.83	0.67	−4.63	−64.34

Note Negative values in percentage error represent lower prediction values than experimental data

nanocomposites had poor filler-matrix interfacial bonding. However, other theoretical models such as D-T model and Lu model presented an overestimation of tensile strengths as compared with experimental data. The percentage errors were found to be over 10% for those two models, as seen in Table 7.7.

On the other hand, with respect to T-P-T model, the curve fitting with experimental data resulted in an empirical constant $B = 1.273$. However, the prediction curve could not well fit the experimental data at the ESO contents ranging from 0 to 60 wt%. Furthermore, in the further prediction analysis, the experimental data at 0 wt% were removed in the prediction calculations. As such, experimental data were only used from 20 to 60 wt% in the subsequent model prediction. Such prediction calculations based on three proposed experimental data were shown in Fig. 7.9. The curve fitting with experimental data (20 to 60 wt% ESO) yielded better agreement with T-P-T model prediction. It was found that the value of B was equal to 7.960. From this T-P-T model prediction ($B = 7.960$), the percentage differences between experimental data and theoretical prediction of tensile strength were 0.67 and 4.63% for bionanocomposites with 20 and 40 wt% ESO, respectively. Moreover, T-P-T model ($B = 7.960$) led to a closer correlation with experimental data, as compared to other proposed models for strength prediction, which was clearly seen from the percentage errors according to Table 7.7.

On the other hand, prediction models for tensile strengths of bionanocomposites with 60 wt% ESO presented a high percentage of the difference, as compared with experimental data (Table 7.7). Proposed theoretical models seemed to underestimate experimental data. It was suggested that the high improvement in tensile strength of bionanocomposites (i.e. 60 wt% ESO) could be attributed to structural factors due to different morphological structures and interfacial bonding between fillers and matrices. Turcsányi et al. [22] noted that structural factors, which could influence stress concentration, are dispersed filler shape and size, spatial distribution of matrices, as well as interface thickness. In practically, as previously mentioned in the theoretical predictions of elastic moduli of bionanocomposites, increasing the ESO content exhibited increasing aspect ratios of clay fillers in matrices. Furthermore, the improvement of aspect ratios can also be associated with the reduction of filler size in the matrices, which could yield a much higher increase in clay surface area. As such, high clay surface areas could cause better matrix-filler interaction, which induced positive impact on the strength improvement resulting in more efficient stress transfer between fillers and matrices.

7.6 Summary

A comparison between experimental data and theoretical predictions of tensile properties of bioepoxy/clay nanocomposites has been shown. Two different tensile property data (i.e. tensile modulus and tensile strength) were used and fitted with proposed models. Theoretical models selected for analysing the tensile modulus were Voigt model and Reuss model, Hirsch model, Halpin–Tsai (H-T) model, modified H-T model, Hua-Shia (H–S) model, as well as laminate model. In addition, theoretical predictions for tensile strengths were conducted by employing Danusso-Tieghi (D-T) model, Nicolais-Narkis (N–N) model, Lu model and Turcsányi-Pukànszky-Tüdõs (T-P-T) model. Tensile properties of randomly oriented clay fillers were presented as a function of clay content and ESO content. Moreover, due to the imperfect filler orientation with the combination of intercalated/exfoliated clay structures, as well as aggregated clay particles, the modulus reduction factor (MRF$\approx$0.66) was used as a constant along with aspect ratios of fillers.

With respect to the theoretical prediction of tensile modulus, all proposed theoretical models were applied to experimental data as a function of clay content and most models demonstrated good agreement with experimental data except for Voigt model and Reuss model with either overestimated or underestimated results, respectively. In addition, Hirsch model with the combination of series and parallel filler orientations revealed a reasonable prediction for constant parameter ("x"), which was equal to 0.029 in the parallel direction. The percentage error was found about 10–20% at all different clay contents. Among various models, H-T laminate model and modified H-T random oriented model with MRF, as well as H–S laminate model showed a very good correlation with experimental data, with the percentage errors being below 3.5%. Nonetheless, it was also exhibited that aspect ratios of clay fillers in bionanocomposites increased with increasing the clay filler content. On the other hand, theoretical predictions of tensile modulus as a function of ESO content demonstrated a partial correlation with experimental data for most proposed models except Voigt model. Similar to the effect of clay content, Voigt model overestimated experimental data, which could be attributed to the assumption of full longitudinal orientation of clay fillers. With respect to conventional nanocomposites (i.e. 0 wt% ESO), lower-bound model predictions (i.e. Reuss model and Hirsch model with $x = 0.006$ for the parallel filler orientation), as well as H–S laminate model fitted well with experimental data with the error percentages being less than 10%. However, at the ESO contents of 20–40 wt%, H-T laminate model, modified H-T random oriented model, as well as H–S laminate model demonstrated very good agreement with experimental data with the percentage errors being less than 5%. In addition, for the experimental data of bionanocompsoites with 60 wt% ESO, most proposed theoretical models underestimated experimental data except Hirsch model. A good agreement between experimental data and Hirsch model was observed when the value of "x" was equal to 0.006 for the parallel filler orientation. This suggested that the series orientation of fillers was more dominant in predicting elastic moduli of bionanocomposites with 60 wt% ESO where the error percentage was about 7%. Additionally, based on H-T model and H–S model, it was also revealed that aspect ratios of clay fillers in nanocomposites increased with increasing the ESO content.

On the other hand, with respect to tensile strength as a function of clay content, all proposed models except N–N model and T-P-T model ($B = 0$), partially fitted well with experimental data. It was influenced by experiment data at the clay content of 8 wt% with a typical declining trend. N–N model and T-P-T model (B = 0) underestimated experiment data because those models were assumed to encounter no load transfer from matrices to fillers due to poor interfacial bonding or no adhesion. Accordingly, the total load was determined based on the load carried by the matrices only. In addition, D-T model, Lu model, as well as T-P-T model (B = 3) had good partial agreement with experimental data with 5 wt% clay inclusion. However, the curve fitting of experimental data at clay contents from 0 to 5 wt% revealed good accordance when T-P-T model was employed with empirical constant B equal to 8.693. This result suggested strong filler-matrix interfacial bonding took place accordingly. Additionally, with respect to the effect of ESO content on tensile strengths of bionanocomposites, selected models except D-T model and Lu model showed good partial fitting with experimental data. N–N model and T-P-T model ($B = 0$) fully fitted only the experimental data of conventional nanocomposites (i.e. 0 wt% ESO). In contrast, T-P-T model fitted well experimental data at the ESO contents of 20 and 40 wt%, which presented strong filler-matrix interfacial bonding with B being equal to 7.960. In addition, experimental data with 60 wt% ESO did not fit all proposed models with a typical underestimation. This suggested that the microstructures (i.e. interfacial structural parameters including the size of dispersed fillers, matrix distribution, as well as thickness of interface area) were not involved in proposed theoretical models.

All proposed theoretical prediction in this study for tensile properties indicated the effect on tensile modulus and tensile strength of bionanocomposites associated with blended matrices, clay filler type, the orientation of clay fillers, filler dispersion status, structure of clay fillers in matrices, as well as filler-matrix interfacial bonding. However, the parameters related to more critical material features of matrix-filler interface, such as thickness, modulus and strength as well as a specific surface areas in the interphase were not specifically investigated here and can be evaluated holistically in the near future.

References

1. Zeng QH, Yu AB, Lu GQ (2008) Multiscale modeling and simulation of polymer nanocomposites. Prog Polym Sci 33(2): 191–269.
2. Voigt W (1889) Ueber die Beziehung zwischen den beiden Elasticitätsconstanten isotroper Körper. Ann Phys 274(12): 573–587.
3. Reuss A (1929) Berechnung der Fließgrenze von Mischkristallen auf Grund der Plastizitätsbedingung für Einkristalle. ZAMM 9(1): 49–58.
4. Fornes TD, Paul DR (2003) Modeling properties of nylon 6/clay nanocomposites using composite theories. Polymer 44(17): 4993–5013.
5. Brune DA, Bicerano J (2002) Micromechanics of nanocomposites: comparison of tensile and compressive elastic moduli, and prediction of effects of incomplete exfoliation and imperfect alignment on modulus. Polymer 43(2):369–387.
6. Hui CY, Shia D (1998) Simple formulae for the effective moduli of unidirectional aligned composites. Polym Eng Sci 38(5): 774–782.
7. Wu Y-P, Jia Q-X, Yu D-S, Zhang L-Q (2004) Modeling Young's modulus of rubber–clay nanocomposites using composite theories. Polym Test 23(8): 903–909.
8. Riley VR (1970) Interaction effects in polymer composites. University of Utah: Salt Lake City.

9. Padawer GE, Beecher N (1970) On the strength and stiffness of planar reinforced plastic resins. Polym Eng Sci 10(3):185–192.
10. Hirsch TJ (1962) Modulus of elasticity of concrete affected by elastic moduli of cement paste matrix and aggregate. J Amer Conc Inst 59 (3): 427–452.
11. Halpin JC (1969) Stiffness and expansion estimates for oriented short fiber composites. J Compos Mater 3(4): 732–734.
12. Affdl JCH, Kardos JL (1976) The Halpin-Tsai equations: A review. Polym Eng Sci 16(5): 344–352.
13. Qian D, Dickey EC, Andrews R, Rantell T (2000). Load transfer and deformation mechanisms in carbon nanotube-polystyrene composites. Appl Phys Lett 76(20): 2868–2870.
14. Mallick PK (2007) Fiber-Reinforce Composites: Materials, Manufacturing, and Design. CRC Press: Boca Raton.
15. Shia D, Hui CY, Burnside SD, Giannelis EP (1998) An interface model for the prediction of Young's modulus of layered silicate-elastomer nanocomposites. Polym Compos 19(5): 608–617.
16. Hull D, Clyne TW (1996) An Introduction to Composite Materials. Cambridge University Press: Cambridge, 78–104.
17. van Es M, Xiqiao F, van Turnhout J, van der Giessen E (2001) Comparing polymer-clay nanocomposites with conventional composites using composite modeling. In: AlMalaika S, Golovoy A, Wilkie CA (Eds) Specialty Polymer Additives: Principles and Applications. Blackwell Science Publishers: Oxford, 391–414.
18. Danusso F, Tieghi G (1986) Strength versus composition of rigid matrix particulate composites. Polymer 27(9):1385–1390.
19. Delesse A (1848) Pour determiner la composition des roches. Annales des Mines 13: 379–388.
20. Nicolais L, Narkis M (1971) Stress-strain behavior of styrene-acrylonitrile/glass bead composites in the glassy region. Polym Eng Sci 11(3):194–199.
21. S Lu, L Yan, X Zhu, Z Qi (1992) Microdamage and interfacial adhesion in glass bead-filled high-density polyethylene. J Mater Sci 27 (17) 4633–4638.
22. Turcsányi B, Pukánszky B, Tüdõs F (1988) Composition dependence of tensile yield stress in filled polymers. J Mater Sci Lett 7(2): 160–162.
23. Salma H, Dong Y (2019) Theoretical modelling analysis on tensile properties of bioepoxy/clay nanocomposites using epoxidised soybean oils. J Nanomater 2019: 4074869.
24. Pavlidou S, Papaspyrides CD (2008) A review on polymer layered silicate nanocomposites. Prog Polym Sci 33(12):1119–1198.
25. Fu S-Y, Feng X-Q, Lauke B, Mai Y-W (2008) Effects of particle size, particle/matrix interface adhesion and particle loading on mechanical properties of particulate–polymer composites. Compos Part B: Eng 39(6):933–961.

Chapter 8
Sustainable Polymer Nanocomposite Applications

8.1 Automotive Applications

In automotive industries, polymer nanocomposite materials can be used in the production of lightweight vehicles to improve fuel efficiency, as well as reduce CO_2 emission. Global expectation on sustainable materials, which are environmentally friendly, lightweight with high material performance and low cost, are attracting researchers to replace metals in automotive industries [1, 2]. Polymer nanocomposites are widely used as material components in exteriors and interiors at transportation sectors, as listed in Table 8.1. Toyota and Ube Companies first pioneered the application of polymer nanocomposites for sophisticated manufactured automobiles in 1990s [3]. They produced automotive components such as timing belts, engines, inverters, and headlamp covers that were made from polyamide 6/clay nanocomposites. Since then, such a promising trend has been followed by other leading global companies such as General Motors, Mitsubishi GDI, and Pirelli, which produced automobile components and tyres from nylon, polyolefins and rubbers, respectively [4]. The key to a commercial success lies in the utilisation of a small amount of exfoliated montmorillonite (MMT) reinforcements in polymer nanocomposites with the cost-effectiveness for the entire industrial-level manufacturing of automotive parts such as timing belt cover, exterior step assist, fuel line, etc.

On the other hand, Nobel et al. [5] focused on waterborne acrylic polymer nanocomposites used for automotive coating with the incorporation of clays such as MMT, laponite and boehmite into organic acrylic polymers. It was reported that aspect ratio, organomodification and nanoparticle location could play an important role in well-tailored mechanical and rheological properties of nanocomposites. In particular, the shielding effect of nanoparticles by means of covered surfaces with polymer matrices appeared to be more beneficial to increase the closer filler-matrix interaction. Notta-Cuvier et al. [6] investigated the durability of polylactic acid (PLA) composites reinforced with halloysite nanotubes (HNTs) for automotive applications. Mechanical properties in terms of strength and rigidity of PLA/HNT composites can be modified by the inclusion of plasticisers in order to obtain desirable composite

H. Salam and Y. Dong, *Bioepoxy/Clay Nanocomposites*,
https://doi.org/10.1007/978-981-16-7297-2_8

Table 8.1 Commercial applications of polymer nanocomposites [4, 7]

Polymer matrix	Nanofillers	Intended properties	Application	Company/Trade name
PA nylon 6, 66, 12	Exfoliated clays	Stiffness, heat resistance, barrier properties, tensile strength, elasticity and impact resistance	Timing belt cover, engine cover, fuel line, fuel hoses/valves, fuel tanks, material packaging, medical devices and flame-retardant materials	Toyota/Ube Industries, Ltd, RTP company, Unika Ltd, Nylon corporation of America (NYCOA, SABIC, Innovatives Plastic Pty. Ltd.) and Ube (Ecobesta®)
TPO	Exfoliated clays	Stiffness/strength and heat resistance	Vehicles step assist, door frame, seta backs, sail panel, box rail, side trim, grills, instrument panels and fenders	General motors (GM) (Chevrolet Impala, Hummer H2), Lyndell Basell, Ube Industries, Ltd., Safari and Astro and DuPont
PP	Clays	Stiffness/strength	Step-assist in mid van	GM
Epoxy	CNTs	Stiffness/strength	Tennis rackets and hockey stick	Babolat, Montreak: Nitro Hybtonite
Ethyl vinyl acetate	MMT-Clay	Flame retardancy, chemical resistance, thermal stability, ease of processing and halogen-free feature	Cable and wire sheeting, as well as semiconductor shields	Exxonmobile (Escorene™ Ultra EVA), Süd-Chemie, Kabelwerk EUPEN AG
Polyisobutylene rubber	Exfoliated clays	Barrier properties	Tire, tennis and soccer ball	InMat LLC (Air D-Fense)
Unknown	Silver	Antimicrobial properties	Bandage	Hyperion
Nylon MXD6,	Exfoliated clays	Barrier properties	Packaging	Curad®
PP	Clays	Gas barrier, high flexural modulus, high impact resistance, low bulk density, mark resistance and fire resistance	Packaging parts, office furniture and automotive parts	Clariant (Durethan), Noble polymer (Forte), PolyOne (Nanoblend)

(continued)

Table 8.1 (continued)

Polymer matrix	Nanofillers	Intended properties	Application	Company/Trade name
Natural rubber	Silver	Wear resistance	Tire	Pirelli S.p.A, Yokohama tyre Corp., The Goodyear tire and rubber Co., Continental AG
Natural rubber	Silver	Antimicrobial	Latex glove	InMat Inc
Nylon	Clays	Good flexibility and barrier properties	Film packaging	Nanocor, Inc. (Imperm® line), Mitsubishi Chemicals, Honeywell (Aegis® line), Bayer AG (Durethan® LPDU), Alcoz CSI
PLA-PET-EVOH	Clays		Meal ready to eat packaging	US Army (Natick Soldier Centre)
PET	Clays		Beer bottle	Nanocor, Inc. (Imperm® line)
PE	Clays	High flexural modulus, high impact resistance, low bulk density, mark resistance and fire resistance	Packaging and automotive parts	InMat (Nanolok), LG Chemicals, PolyOne (Maxxam LST)
Nylon 12	Clays		Medical devices	Foster Corp. (Nanomed)

Note PA: polyamides; TPO: thermoplastic polyolefin; PP: Polypropylene; CNTs: carbon nanotubes; MMT: montmorillonite; EVA (or EVOH): ethylene–vinyl alcohol; PET: polyethylene terephthalate; PE: polyethylene

properties in automotive industries such as impact resistance, ductility, degree of crystallinity, etc. It was identified in this study that an optimal composition of PLA/HNT composites containing 10 wt% tributyl citrate (TBC) as an effective plasticiser could become a potential biosourced alternative for automotive application.

8.2 Material Packaging

In food packaging industries, polymeric packaging materials are most widely used due to their cost-effectiveness and easy processability as compared to other types of packaging materials based on metals, glass and paper [8]. Silvestre et al. [9] reviewed food packaging applications based on polymer nanocomposites. Material packaging can be improved in term of the effect of dispersed nanoclays into polymer matrices. The additional clay nanofillers are not only enhancing mechanical properties but also improving oxygen and water barrier properties, antimicrobial properties, as well as humidity and temperature resistance, which would intensify the development of polymer nanocomposites in packaging sectors [8]. Lagaron and Núñez [10] investigated the effect of clay content and humidity condition on water and gas barrier properties of ethylene–vinyl alcohol (EVA) copolymer nanocomposites. The additional 4 wt% clays were shown to reduce water and gas permeabilities by 42 and 29%, respectively. In addition, increasing relative humidity condition up to 90% was found to cause the reduction in oxygen permeability by 71%.

On the other hand, nanocomposites as active packaging materials denote those with active functions, which are considered as food preservatives on protecting the products from the surrounding environment, and are used to absorb or release substances into or from the packaging [8, 11]. Such active packaging features of sustainable polymer nanocomposites could be used for antimicrobial and antioxidant packaging applications. Rhim et al. [12] prepared PLA /clay nanocomposite films by solvent casting. The effect of different nanoclay types, namely Cloisite Na^+, Cloisite 20A and Cloisite 30B clays on antimicrobial properties was investigated. Among various nanoclay types, PLA/Cloisite 30B clay nanocomposites showed better antimicrobial characteristic against *Listeria monocytogenes*. Similarly, Shin et al. [13] investigated antimicrobial properties of barley protein/clay nanocomposites. It was confirmed that barley protein/Cloisite Na^+ nanocomposite films containing grapefruit extracts could be potentially used for antimicrobial packaging of high-quality mushroom during food and vegetable storage, and further considered as edible food packaging films in food industries.

The potential use of sustainable polymer nanocomposites can provide the necessity for modern packaging industries. Moreover, the incorporation of clay nanofillers into polymer matrices can improve mechanical, oxygen and water barrier properties, as well as antimicrobial properties [8]. In addition, it is also noted that the applications of those nanocomposites in food packaging sectors must also fulfil the requirements for sustainability, product quality, convenience and food safety [8].

8.3 Medical Applications

Notwithstanding biocompatibility and functionalisation of biopolymers, their corresponding bionanocomposites make it possible to be adapted for medical applications. Currently, the application of biopolymers and their bionanocomposites in medical sectors can be divided into two aspects of areas, namely tissue engineering and drug delivery. With respect to tissue engineering, the applications of biopolymers and their bionanocomposites tend to create artificial tissues as scaffolding materials [14, 15]. Tissue engineering can be employed as the guidance to cell outgrowth and nerve regeneration, vascular grafts, bone regeneration, as well as skin tissue engineering in order to regenerate or replace deteriorating tissues [16, 17]. Among several biopolymers, saturated poly (α-hydroxy esters) (i.e. PLA, poly(glycol acid) (PGA), poly(lactic acid-co-glycolic) (PLGA) and poly(ε-caprolactone) (PCL) are widely used as tissue scaffolds [14, 18, 19]. Additionally, the functionalisation of clay fillers (i.e. MMTs and HNTs) into biopolymer matrices has also been potentially used in tissue engineering such as polyvinyl alcohol (PVA)/HNT bionanocomposites as bone tissues [20], chitosan/HNT nanocomposites as tissue scaffolding materials [21, 22] and poly(l-lactic acid) (PLLA)/MMT nanocomposites that can be implemented for the invasion of blood vessels and facial cell growth [23].

On the other hand, the stabilisation of label molecules from degradation and specific drug target can also benefit from the utilisation of biopolymers and their nanocomposites as drug release agents. Generally, the drugs can be dissolved, adsorbed, attached and/or encapsulated into nanocomposites. In addition, nanoclay fillers with the diameters in a size range of 1–1000 nm enable the drugs in solutions to penetrate into interlayer areas via ion exchange reaction. Moreover, in nanocomposites systems, drugs can be adsorbed on nanocomposite surfaces by dipping them in the active solution. Nanocomposites reinforced with MMT and HNT nanofillers have been identified to be applicable as a useful and safe material in drug delivery [24–26]. In particular, Viseras et al. [26] pinpointed that the use of polymer/clay bionanocomposites could shed a light on the extended application of designing new dosage forms of drug release with highly specific technological and biopharmaceutical characteristics in terms of swelling, film forming, bioadhesion and cell uptake, which was based upon the remarkable enhancements of increasing mechanical and rheological properties, reducing drug permeability and water uptake, as well as controlling drug release. Banik et al. [27] evaluated the effect of glutaraldehyde and MMTs of carboxymethyl chitosan (CMC)-MMT nanoparticles on isoniazid delivery. By decreasing the MMT and glutaraldehyde contents, the swelling and release of isoniazid from clay nanoparticles increased. Moreover, the cytotoxic level of CMC-MMT nanoparticles was observed to be less as opposed to MMT particles. Similarly, Wang et al. [28] studied the effect of combined chitosan and MMT nanocomposites on drug encapsulation efficiency. It was suggested that a certain amount of intercalated MMT could enhance drug encapsulation effectively, reduce drug release from nanocomposites and improve noncytotoxic level thanks to the positive effect of MMT particles relating to drug adsorption action and mucosa protection.

8.4 Summary

The widespread use of sustainable polymer nanocomposites in automotive, material packaging and biomedical applications relies primarily on biodegradable, biocompatible or non-toxic polymeric materials as the matrix when incorporated with a small amounts of well-dispersed and environmentally friendly or user-friendly nanofillers. Clearly, sophisticated manufacturing process with cost-effectiveness, well-tailored structures by means of uniform nanofiller dispersion and significantly improved material properties of nanocomposites are the key to the commercial success of such advanced materials at a larger industrial level.

References

1. Komarneni S (1992) Nanocomposites. J Mater Chem 2(12): 1219–1230.
2. Garcés JM, Moll DJ, Bicerano J, Fibiger R, McLeod DG (2000) Polymeric nanocomposites for automotive applications. Adv Mater 12(23):1835–1839.
3. Usuki A, Kojima Y, Kawasumi M, Okada A, Fukushima Y, Kurauchi T, Kamigaito O (1993) Synthesis of nylon 6-clay hybrid. J Mater Res 8(5): 1179–1184.
4. Paul DR, Robeson LM (2008) Polymer nanotechnology: Nanocomposites. Polymer 49(15): 3187–3204.
5. Nobel ML, Picken SJ, Mendes E (2007) Waterborne nanocomposite resins for automotive coating applications. Prog Org Coat 58(2–3): 96–104.
6. Notta-Cuvier D, Murariu M, Odent J, Delille R, Bouzouita A, Raquez J-M, Lauro F, Dubois P (2015). Tailoring polylactide properties for automotive applications: Effects of co-addition of halloysite nanotubes and selected plasticiser. Macromol Mater Eng 300(7): 684–698.
7. Ray SS (2013) Clay-Containing Polymer Nanocomposites. Elsevier: Amsterdam, 369–373.
8. Youssef AM (2013) Polymer nanocomposites as a new trend for packaging applications. Polym-Plast Technol Eng 52(7): 635–660.
9. Silvestre C, Duraccio D, Cimmino S (2011) Food packaging based on polymer nanomaterials. Prog Polym Sci 36(12):1766–1782.
10. Lagaron JM, Núñez E (2012) Nanocomposites of moisture-sensitive polymers and biopolymers with enhanced performance for flexible packaging applications. J Plast Film Sheeting 28(1): 79–89.
11. Jamshidian M, Tehrany EA, Imran M, Jacquot M, Desobry S (2010) Poly-lactic acid: Production, applications, nanocomposites, and release studies. Compr Rev Food Sci 9(5): 552–571.
12. Rhim J-W, Hong S-I, Ha C-S (2009) Tensile, water vapor barrier and antimicrobial properties of PLA/nanoclay composite films. LWT–Food Sci Technol 42(2): 612–617.
13. Shin Y-J, Song H-Y, Jo W-S, Lee M-J, Song KB (2013) Physical properties of a barley protein/nano-clay composite film containing grapefruit seed extract and antimicrobial benefits for packaging of Agaricus bisporus. Inter J Food Sci Technol 48(8):1736–1743.
14. Okamoto M, John B (2013) Synthetic biopolymer nanocomposites for tissue engineering scaffolds. Prog Polym Sci 38(10–11):1487–1503.
15. Fernandes EM, Pires RA, Mano JF, Reis RL (2013) Bionanocomposites from lignocellulosic resources: Properties, applications and future trends for their use in the biomedical field. Prog Polym Sci 38(10–11): 1415–1441.
16. Sun H-Y, Yin T, Zhao, J-B, Zhang Z-Y, Yang W-T (2013) Synthesis and characterisation of biodegradable polyesteramides constructed mainly by alternating diesterdiamide units from N,N′-bis(2-hydroxyethyl)-adipamide and diacids. Chinese J Polym Sci 31(3): 452–461.
17. Fraczek-Szczypta A (2014) Carbon nanomaterials for nerve tissue stimulation and regeneration. Mater Sci Eng C 34: 35–49.
18. Woodruff MA, Hutmacher DW (2010) The return of a forgotten polymer—Polycaprolactone in the 21st century. Prog Polym Sci 35(10):1217–1256.
19. Armentano I, Bitinis N, Fortunati E, Mattioli S, Rescignano N, Verdejo R, Lpoez-Manchado MA, Kenny JM (2013) Multifunctional nanostructured PLA materials for packaging and tissue engineering. Prog Polym Sci 38(10–11):1720–1747.
20. Zhou WY, Guo B, Liu M, Liao R, Rabie ABM, Jia D (2010) Poly(vinyl alcohol)/halloysite nanotubes bionanocomposite films: Properties and in vitro osteoblasts and fibroblasts response. J Biomed Mater Res A 93A(4): 1574–1587.

21. Liu M, Guo B, Du M, Lei Y, Jia D (2008) Natural inorganic nanotubes reinforced epoxy resin nanocomposites. J Polym Res 15(3) 205–212.
22. Croisier F, Jérôme C (2013) Chitosan-based biomaterials for tissue engineering. Eur Polym J 49(4): 780–792.
23. Lee YH, Lee JH, An I-G, Kim C, Lee DS, Lee YK, Nam J-D (2005) Electrospun dual-porosity structure and biodegradation morphology of montmorillonite reinforced PLLA nanocomposite scaffolds. Biomaterials 26(16): 3165–3172.
24. Levis SR, Deasy PB (2002) Characterisation of halloysite for use as a microtubular drug delivery system. Inter J Pharm 243(1–2):125–134.
25. Aguzzi C, Cerezo P, Viseras C, Caramella C (2007) Use of clays as drug delivery systems: Possibilities and limitations. Appl Clay Sci 36(1–3): 22–36.
26. Viseras C, Aguzzi C, Cerezo P, Bedmar MC (2008) Biopolymer–clay nanocomposites for controlled drug delivery. Mater Sci Technol 24(9):1020–1026.
27. Banik N, Ramteke A, Maji TK (2014) Carboxymethyl chitosan-montmorillonite nanoparticles for controlled delivery of isoniazid: evaluation of the effect of the glutaraldehyde and montmorillonite. Polym Adv Technol 25(12):1580–1589.
28. Wang X, Du Y, Lu J (2008) Biopolymer/montmorillonite nanocomposite: preparation, drug-controlled release property and cytotoxicity. Nanotechnology 19(6): 065707.

The manufacturer's authorised representative in the EU is Springer Nature Customer Service Centre GmbH, Europaplatz 3, 69115 Heidelberg, Germany. If you have any concerns regarding our products, please contact ProductSafety@springernature.com

Printed and bound by CPI Group (UK) Ltd, Croydon, CR0 4YY
30/09/2026
02225508-0004